1983

# GEOMETRIC EXERCISES
# IN PAPER FOLDING

## by T. Sundara Row

Edited and Revised by

WOOSTER WOODRUFF BEMAN

and

DAVID EUGENE SMITH

DOVER PUBLICATIONS, INC., NEW YORK

This Dover edition, first published in 1966, is an
unabridged and unaltered republication of the second edition, as published by The Open Court Publishing Company in 1905.

*International Standard Book Number: 0-486-21594-6*
*Library of Congress Catalog Card Number: 66-19051*

Manufactured in the United States of America
Dover Publications, Inc.
180 Varick Street
New York, N. Y. 10014

# CONTENTS.

# EDITORS' PREFACE.

Our attention was first attracted to Sundara Row's *Geometrical Exercises in Paper Folding* by a reference in Klein's *Vorlesungen über ausgewählte Fragen der Elementargeometrie*. An examination of the book, obtained after many vexatious delays, convinced us of its undoubted merits and of its probable value to American teachers and students of geometry. Accordingly we sought permission of the author to bring out an edition in this country, which permission was most generously granted.

The purpose of the book is so fully set forth in the author's introduction that we need only to say that it is sure to prove of interest to every wide-awake teacher of geometry from the graded school to the college. The methods are so novel and the results so easily reached that they cannot fail to awaken enthusiasm.

Our work as editors in this revision has been confined to some slight modifications of the proofs, some additions in the way of references, and the insertion of a considerable number of half-tone reproductions of actual photographs instead of the line-drawings of the original.

W. W. Beman.
D. E. Smith.

# INTRODUCTION.

THE *idea* of this book was suggested to me by Kindergarten Gift No. VIII.—Paper-folding. The gift consists of two hundred variously colored squares of paper, a folder, and diagrams and instructions for folding. The paper is colored and glazed on one side. The paper may, however, be of self-color, alike on both sides. In fact, any paper of moderate thickness will answer the purpose, but colored paper shows the creases better, and is more attractive. The kindergarten gift is sold by any dealers in school supplies; but colored paper of both sorts can be had from stationery dealers. Any sheet of paper can be cut into a square as explained in the opening articles of this book, but it is neat and convenient to have the squares ready cut.

2. These exercises do not require mathematical instruments, the only things necessary being a penknife and scraps of paper, the latter being used for setting off equal lengths. The squares are themselves simple substitutes for a straight edge and a T square.

3. In paper-folding several important geometric processes can be effected much more easily than with

a pair of compasses and ruler, the only instruments the use of which is sanctioned in Euclidean geometry; for example, to divide straight lines and angles into two or more equal parts, to draw perpendiculars and parallels to straight lines. It is, however, not possible in paper-folding to describe a circle, but a number of points on a circle, as well as other curves, may be obtained by other methods. These exercises do not consist merely of drawing geometric figures involving straight lines in the ordinary way, and folding upon them, but they require an intelligent application of the simple processes peculiarly adapted to paper-folding. This will be apparent at the very commencement of this book.

4. The use of the kindergarten gifts not only affords interesting occupations to boys and girls, but also prepares their minds for the appreciation of science and art. Conversely the teaching of science and art later on can be made interesting and based upon proper foundations by reference to kindergarten occupations. This is particularly the case with geometry, which forms the basis of every science and art. The teaching of plane geometry in schools can be made very interesting by the free use of the kindergarten gifts. It would be perfectly legitimate to require pupils to fold the diagrams with paper. This would give them neat and accurate figures, and impress the truth of the propositions forcibly on their minds. It would not be necessary to take any statement on trust.

But what is now realised by the imagination and ideal-
isation of clumsy figures can be seen in the concrete.
A fallacy like the following would be impossible.

5. *To prove that every triangle is isosceles.* ' Let
*ABC*, Fig. 1, be any triangle. Bisect *AB* in *Z*, and
through *Z* draw *ZO* perpendicular to *AB*. Bisect the
angle *ACB* by *CO*.

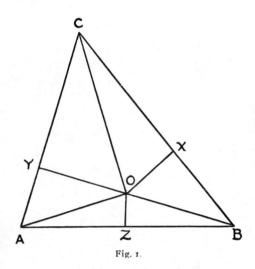

Fig. 1.

(1) If *CO* and *ZO* do not meet, they are parallel.
Therefore *CO* is at right angles to *AB*. Therefore
*AC = BC*.

(2) If *CO* and *ZO* do meet, let them meet in *O*.
Draw *OX* perpendicular to *BC* and *OY* perpendicular
to *AC*. Join *OA*, *OB*. By Euclid I, 26 (B. and S.,
§ 88, cor. 7)* the triangles *YOC* and *XOC* are con-

---

*These references are to Beman and Smith's *New Plane and Solid Geom-
etry*, Boston, Ginn & Co., 1899.

gruent; also by Euclid I, 47 and I, 8 (B. and S., § 156 and § 79) the triangles $AOY$ and $BOX$ are congruent. Therefore

$$AY + YC = BX + XC,$$
$$\text{i. e., } AC = BC.$$

Fig. 2 shows by paper-folding that, whatever triangle be taken, $CO$ and $ZO$ cannot meet within the triangle.

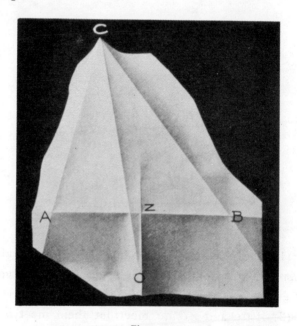

Fig. 2.

$O$ is the mid-point of the arc $AOB$ of the circle which circumscribes the triangle $ABC$.

6. Paper-folding is not quite foreign to us. Folding paper squares into natural objects—a boat, double

boat, ink bottle, cup-plate, etc., is well known, as also the cutting of paper in symmetric forms for purposes of decoration. In writing Sanskrit and Mahrati, the paper is folded vertically or horizontally to keep the lines and columns straight. In copying letters in public offices an even margin is secured by folding the paper vertically. Rectangular pieces of paper folded double have generally been used for writing, and before the introduction of machine-cut letter paper and envelopes of various sizes, sheets of convenient size were cut by folding and tearing larger sheets, and the second half of the paper was folded into an envelope inclosing the first half. This latter process saved paper and had the obvious advantage of securing the post marks on the paper written upon. Paper-folding has been resorted to in teaching the XIth Book of Euclid, which deals with figures of three dimensions.* But it has seldom been used in respect of plane figures.

7. I have attempted not to write a complete treatise or text-book on geometry, but to show how regular polygons, circles and other curves can be folded or pricked on paper. I have taken the opportunity to introduce to the reader some well known problems of ancient and modern geometry, and to show how algebra and trigonometry may be advantageously applied to geometry, so as to elucidate each of the subjects which are usually kept in separate pigeon-holes.

* See especially Beman and Smith's *New Plane and Solid Geometry*, p. 287.

8. The first nine chapters deal with the folding of the regular polygons treated in the first four books of Euclid, and of the nonagon. The paper square of the kindergarten has been taken as the foundation, and the other regular polygons have been worked out thereon. Chapter I shows how the fundamental square is to be cut and how it can be folded into equal right-angled isosceles triangles and squares. Chapter II deals with the equilateral triangle described on one of the sides of the square. Chapter III is devoted to the Pythagorean theorem (B. and S., § 156) and the propositions of the second book of Euclid and certain puzzles connected therewith. It is also shown how a right-angled triangle with a given altitude can be described on a given base. This is tantamount to finding points on a circle with a given diameter.

9. Chapter X deals with the arithmetic, geometric, and harmonic progressions and the summation of certain arithmetic series. In treating of the progressions, lines whose lengths form a progressive series are obtained. A rectangular piece of paper chequered into squares exemplifies an arithmetic series. For the geometric the properties of the right-angled triangle, that the altitude from the right angle is a mean proportional between the segments of the hypotenuse (B. and S., § 270), and that either side is a mean proportional between its projection on the hypotenuse and the hypotenuse, are made use of. In this connexion the Delian problem of duplicating a cube has been

explained.* In treating of harmonic progression, the fact that the bisectors of an interior and corresponding exterior angle of a triangle divide the opposite side in the ratio of the other sides of the triangle (B. and S., § 249) has been used. This affords an interesting method of graphically explaining systems in involution. The sums of the natural numbers and of their cubes have been obtained graphically, and the sums of certain other series have been deduced therefrom.

10. Chapter XI deals with the general theory of regular polygons, and the calculation of the numerical value of $\pi$. The propositions in this chapter are very interesting.

11. Chapter XII explains certain general principles, which have been made use of in the preceding chapters,—congruence, symmetry, and similarity of figures, concurrence of straight lines, and collinearity of points are touched upon.

12. Chapters XIII and XIV deal with the conic sections and other interesting curves. As regards the circle, its harmonic properties among others are treated. The theories of inversion and co-axial circles are also explained. As regards other curves it is shown how they can be marked on paper by paper-folding. The history of some of the curves is given, and it is shown how they were utilised in the solution

---

*See Beman and Smith's translation of Klein's *Famous Problems of Elementary Geometry*, Boston, 1897; also their translation of Fink's *History of Mathematics*, Chicago, The Open Court Pub. Co., 1900.

of the classical problems, to find two geometric means between two given lines, and to trisect a given rectilineal angle. Although the investigation of the properties of the curves involves a knowledge of advanced mathematics, their genesis is easily understood and is interesting.

13. I have sought not only to aid the teaching of geometry in schools and colleges, but also to afford mathematical recreation to young and old, in an attractive and cheap form. "Old boys" like myself may find the book useful to revive their old lessons, and to have a peep into modern developments which, although very interesting and instructive, have been ignored by university teachers.

T. Sundara Row.

Madras, India, 1893.

# I. THE SQUARE.

**1.** The upper side of a piece of paper lying flat upon a table is a plane surface, and so is the lower side which is in contact with the table.

**2.** The two surfaces are separated by the material of the paper. The material being very thin, the other sides of the paper do not present appreciably broad surfaces, and the edges of the paper are practically lines. The two surfaces though distinct are inseparable from each other.

**3.** Look at the irregularly shaped piece of paper shown in Fig. 3, and at this page which is rectangular. Let us try and shape the former paper like the latter.

**4.** Place the irregularly shaped piece of paper upon the table, and fold it flat upon itself. Let $X'X$ be the crease thus formed. It is straight. Now pass a knife along the fold und separate the smaller piece. We thus obtain one straight edge.

**5.** Fold the paper again as before along $BY$, so that the edge $X'X$ is doubled upon itself. Unfolding the paper, we see that the crease $BY$ is at right angles to the edge $X'X$. It is evident by superposition that

the angle $YBX'$ equals the angle $XBY$, and that each
of these angles equals an angle of the page. Now pass

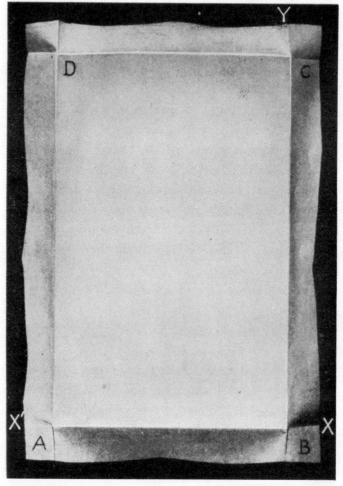

Fig. 3.

a knife as before along the second fold and remove
the smaller piece.

**6.** Repeat the above process and obtain the edges *CD* and *DA*. It is evident by superposition that the angles at *A*, *B*, *C*, *D*, are right angles, equal to one another, and that the sides *BC*, *CD* are respectively

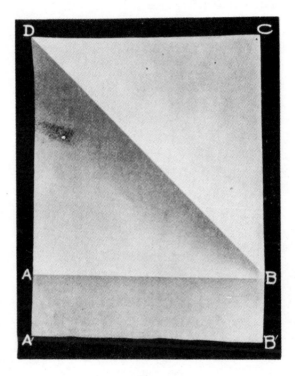

Fig. 4.

equal to *DA*, *AB*. This piece of paper (Fig. 3) is similar in shape to the page.

**7.** It can be made equal in size to the page by taking a larger piece of paper and measuring off *AB* and *BC* equal to the sides of the latter.

**8.** A figure like this is called a rectangle. By superposition it is proved that (1) the four angles are right angles and all equal, (2) the four sides are not all equal, (3) but the two long sides are equal, and so also are the two short sides.

**9.** Now take a rectangular piece of paper, $A'B'CD$, and fold it obliquely so that one of the short sides, $CD$,

Fig. 5.

falls upon one of the longer sides, $DA'$, as in Fig. 4. Then fold and remove the portion $A'B'BA$ which overlaps. Unfolding the sheet, we find that $ABCD$ is now square, i. e., its four angles are right angles, and all its sides are equal.

**10.** The crease which passes through a pair of the

opposite corners *B*, *D*, is a diagonal of the square. One other diagonal is obtained by folding the square through the other pair of corners as in Fig. 5.

11. We see that the diagonals are at right angles to each other, and that they bisect each other.

12. The point of intersection of the diagonals is called the center of the square.

Fig. 6.

13. Each diagonal divides the square into two congruent right-angled isosceles triangles, whose vertices are at opposite corners.

14. The two diagonals together divide the square into four congruent right-angled isosceles triangles, whose vertices are at the center of the square.

**15.** Now fold again, as in Fig. 6, laying one side of the square upon its opposite side. We get a crease which passes through the center of the square. It is at right angles to the other sides and (1) bisects them; (2) it is also parallel to the first two sides; (3) it is itself bisected at the center; (4) it divides the square

Fig. 7.

into two congruent rectangles, which are, therefore, each half of it; (5) each of these rectangles is equal to one of the triangles into which either diagonal divides the square.

**16.** Let us fold the square again, laying the remaining two sides one upon the other. The crease now obtained and the one referred to in § 15 divide the square into four congruent squares.

17. Folding again through the corners of the smaller squares which are at the centers of the sides of the larger square, we obtain a square which is inscribed in the latter. (Fig. 7.)

Fig. 8.

18. This square is half the larger square, and has the same center.

19. By joining the mid-points of the sides of the inner square, we obtain a square which is one-fourth of the original square (Fig. 8). By repeating the process, we can obtain any number of squares which are to one another as

$$\frac{1}{2}, \ \frac{1}{4}, \ \frac{1}{8}, \ \frac{1}{16}, \ \text{etc.,} \ \text{or} \ \frac{1}{2}, \ \frac{1}{2^2}, \ \frac{1}{2^3}, \ \frac{1}{2^4}, \ \cdots$$

Each square is half of the next larger square, i. e., the four triangles cut from each square are together equal to half of it. The sums of all these triangles increased to any number cannot exceed the original square, and they must eventually absorb the whole of it.

Therefore $\frac{1}{2} + \frac{1}{2^2} + \frac{1}{2^3} +$ etc. to infinity $= 1$.

**20.** The center of the square is the center of its circumscribed and inscribed circles. The latter circle touches the sides at their mid-points, as these are nearer to the center than any other points on the sides.

**21.** Any crease through the center of the square divides it into two trapezoids which are congruent. A second crease through the center at right angles to the first divides the square into four congruent quadrilaterals, of which two opposite angles are right angles. The quadrilaterals are concyclic, i. e., the vertices of each lie in a circumference.

## II. THE EQUILATERAL TRIANGLE.

**22.** Now take this square piece of paper (Fig. 9), and fold it double, laying two opposite edges one upon the other. We obtain a crease which passes through

Fig. 9.

the mid-points of the remaining sides and is at right angles to those sides. Take any point on this line, fold through it and the two corners of the square which are on each side of it. We thus get isosceles triangles standing on a side of the square.

**23.** The middle line divides the isosceles triangle into two congruent right-angled triangles.

**24.** The vertical angle is bisected.

**25.** If we so take the point on the middle line, that

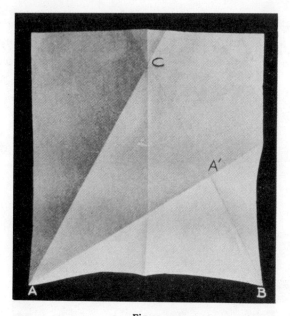

Fig. 10.

its distances from two corners of the square are equal to a side of it, we shall obtain an equilateral triangle (Fig. 10). This point is easily determined by turning the base $AB$ through one end of it, over $AA'$, until the other end, $B$, rests upon the middle line, as at $C$.

**26.** Fold the equilateral triangle by laying each

of the sides upon the base. We thus obtain the three altitudes of the triangle, viz.: $AA'$, $BB'$, $CC'$, (Fig. 11).

**27.** Each of the altitudes divides the triangle into two congruent right-angled triangles.

**28.** They bisect the sides at right angles.

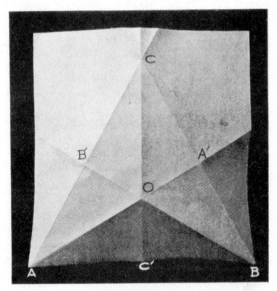

Fig. 11.

**29.** They pass through a common point.

**30.** Let the altitudes $AA'$ and $CC'$ meet in $O$. Draw $BO$ and produce it to meet $AC$ in $B'$. $BB'$ will now be proved to be the third altitude. From the triangles $C'OA$ and $COA'$, $OC' = OA'$. From triangles $OC'B$ and $A'OB$, $\angle OBC' = \angle A'BO$. Again from triangles $ABB'$ and $CB'B$, $\angle AB'B = \angle BB'C$,

i. e., each of them is a right angle.   That is, $BOB'$ is an altitude of the equilateral triangle $ABC$.   It also bisects $AC$ in $B'$.

**31.** It can be proved as above that $OA$, $OB$, and $OC$ are equal, and that $OA'$, $OB'$, and $OC'$ are also equal.

**32.** Circles can therefore be described with $O$ as a center and passing respectively through $A$, $B$, and $C$ and through $A'$, $B'$, and $C$.   The latter circle touches the sides of the triangle.

**33.** The equilateral triangle $ABC$ is divided into six congruent right-angled triangles which have one set of their equal angles at $O$, and into three congruent, symmetric, concyclic quadrilaterals.

**34.** The triangle $AOC$ is double the triangle $A'OC$; therefore, $AO = 2OA'$.   Similarly, $BO = 2OB'$ and $CO = 2OC'$.   Hence the radius of the circumscribed circle of triangle $ABC$ is twice the radius of the inscribed circle.

**35.** The right angle $A$, of the square, is trisected by the straight lines $AO$, $AC$.   Angle $BAC = \frac{2}{3}$ of a right angle.   The angles $C'AO$ and $OAB'$ are each $\frac{1}{3}$ of a right angle.  Similarly with the angles at $B$ and $C$.

**36.** The six angles at $O$ are each $\frac{2}{3}$ of a right angle.

**37.** Fold through $A'B'$, $B'C'$, and $C'A'$ (Fig. 12). Then $A'B'C'$ is an equilateral triangle.   It is a fourth of the triangle $ABC$.

**38.** $A'B'$, $B'C'$, $C'A'$ are each parallel to $AB$, $BC$, $CA$, and halves of them.

**39.** $AC'A'B'$ is a rhombus. So are $C'BA'B'$ and $CB'C'A'$.

**40.** $A'B'$, $B'C'$, $C'A'$ bisect the corresponding altitudes.

**41.** $CC'^2 + AC'^2 = CC'^2 + \frac{1}{4}AC^2 = AC^2$

$\therefore CC'^2 = \frac{3}{4}AC^2$

$\therefore CC' = \frac{1}{2}\sqrt{3} \cdot AC = \frac{1}{2}\sqrt{3} \cdot AB$

$= 0.866\ldots \times AB.$

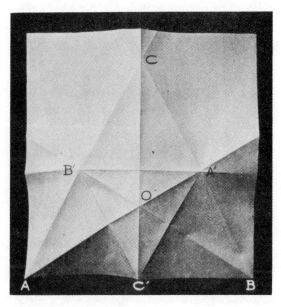

Fig. 12.

**42.** The $\triangle ABC =$ rectangle of $AC'$ and $CC'$, i. e. $\frac{1}{2}AB \times \frac{1}{2}\sqrt{3} \cdot AB = \frac{1}{4}\sqrt{3} \cdot AB^2 = 0.433\ldots \times AB^2.$

**43.** The angles of the triangle $AC'C$ are in the ratio of $1 : 2 : 3$, and its sides are in the ratio of $\sqrt{1} : \sqrt{3} : \sqrt{4}.$

# III. SQUARES AND RECTANGLES.

**44.** Fold the given square as in Fig. 13. This affords the well-known proof of the Pythagorean the-

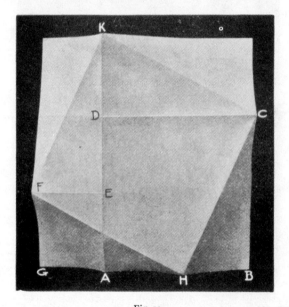

Fig. 13.

orem. $FGH$ being a right-angled triangle, the square on $FH$ equals the sum of the squares on $FG$ and $GH$.

$$\square\, FA + \square\, DB = \square\, FC.$$

It is easily proved that $FC$ is a square, and that

the triangles *FGH*, *HBC*, *KDC*, and *FEK* are congruent.

If the triangles *FGH* and *HBC* are cut off from the squares *FA* and *DB*, and placed upon the other two triangles, the square *FHCK* is made up.

If $AB = a$, $GA = b$, and $FH = c$, then $a^2 + b^2 = c^2$.

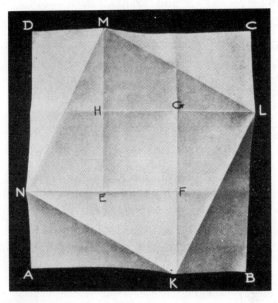

Fig. 14.

**45.** Fold the given square as in Fig. 14. Here the rectangles *AF*, *BG*, *CH*, and *DE* are congruent, as also the triangles of which they are composed. *EFGH* is a square as also *KLMN*.

Let $AK = a$, $KB = b$, and $NK = c$,

then $a^2 + b^2 = c^2$, i. e. □ *KLMN*.

$$\square\, ABCD = (a+b)^2.$$

Now square $ABCD$ overlaps the square $KLMN$ by the four triangles $AKN$, $BLK$, $CML$, and $DNM$.

But these four triangles are together equal to two of the rectangles, i. e., to $2ab$.

Therefore $(a+b)^2 = a^2 + b^2 + 2ab$.

**46.** $EF = a - b$, and $\square\, EFGH = (a-b)^2$.

The square $EFGH$ is less than the square $KLMN$ by the four triangles $FNK$, $GKL$, $HLM$, and $EMN$.

But these four triangles make up two of the rectangles, i. e., $2ab$.

$$\therefore\ (a-b)^2 = a^2 + b^2 - 2ab.$$

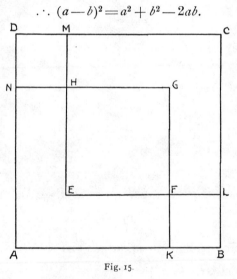

Fig. 15.

**47.** The square $ABCD$ overlaps the square $EFGH$ by the four rectangles $AF$, $BG$, $CH$, and $DE$.

$$\therefore\ (a+b)^2 - (a-b)^2 = 4ab.$$

**48.** In Fig. 15, the square $ABCD = (a+b)^2$, and

the square $EFGH=(a-b)^2$.  Also square $AKGN$ = square $ELCM = a^2$.   Square $KBLF$ = square $NHMD = b^2$.

Squares $ABCD$ and $EFGH$ are together equal to the latter four squares put together, or to twice the square $AKGN$ and twice the square $KBLF$, that is, $(a+b)^2 + (a-b)^2 = 2a^2 + 2b^2$.

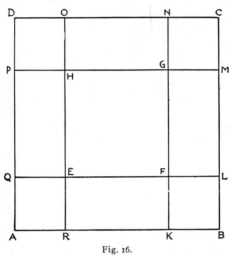

Fig. 16.

**49.** In Fig. 16 the rectangle $PL$ is equal to $(a+b)(a-b)$.

Because the rectangle $EK = FM$, therefore rectangle $PL$ = square $PK$ — square $AE$, i. e., $(a+b)(a-b)=a^2-b^2$.

**50.** If squares be described about the diagonal of a given square, the right angle at one corner being common to them, the lines which join this corner with the mid-points of the opposite sides of the given square bisect the corresponding sides of all the inner squares. (Fig. 17.) For the angles which these lines make with the diagonal are equal, and their magnitude is constant for all squares, as may be seen by

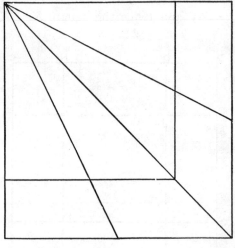

Fig. 17.

superposition. Therefore the mid-points of the sides of the inner squares must lie on these lines.

**51.** *ABCD* being the given square piece of paper (Fig. 18), it is required to obtain by folding, the point *X* in *AB*, such that the rectangle *AB·XB* is equal to the square on *AX*.

Double *BC* upon itself and take its mid-point *E*. Fold through *E* and *A*.

Lay *EB* upon *EA* and fold so as to get *EF*, and *G* such that *EG = EB*.

Take *AX = AG*.

Then rectangle $AB \cdot XB = AX^2$.

Complete the rectangle $BCHX$ and the square $AXKL$.

Let $XH$ cut $EA$ in $M$. Take $FY = FB$.

Then $FB = FG = FY = XM$ and $XM = \frac{1}{2}AX$.

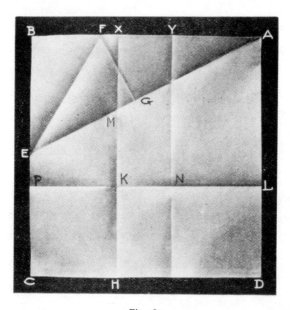

Fig. 18.

Now, because $BY$ is bisected in $F$ and produced to $A$,

$$AB \cdot AY + FY^2 = AF^2, \text{ by } \S 49,$$
$$= AG^2 + FG^2, \text{ by } \S 44.$$
$$\therefore AB \cdot AY = AG^2,$$
$$= AX^2.$$

But $AX^2 = 4 \cdot XM^2 = BY^2$.

$$\therefore AX = BY \text{ and } AY = XB.$$
$$\therefore AB \cdot XB = AX^2.$$

$AB$ is said to be divided in $X$ in median section.*
Also
$$AB \cdot AY = BY^2,$$

i. e., $AB$ is also divided in $Y$ in median section.

**52.** A circle can be described with $F$ as a center, its circumference passing through $B$, $G$, and $Y$. It will touch $EA$ at $G$, because $FG$ is the shortest distance from $F$ to the line $EGA$.

**53.** Since
$$BH = BN,$$
subtracting $BK$ we have
$$\text{rectangle } XKNY = \text{square } CHKP,$$
$$\text{i. e., } AX \cdot YX = AY^2,$$

i. e., $AX$ is divided in $Y$ in median section.

Similarly $BY$ is divided in $X$ in median section.

**54.** $\because AB \cdot XB = AX^2$
$$\therefore 3AB \cdot XB = AX^2 + BX \cdot BC + CD \cdot CP$$
$$= AB^2 + BX^2.$$

**55.** Rectangles $BH$ and $YD$ being each $= AB \cdot XB$, rectangle $HY +$ square $CK = AX^2 = AB \cdot XB$.

**56.** Hence rectangle $HY =$ rectangle $BK$, i. e.,
$AX \cdot XB = AB \cdot XY$.

**57.** Hence rectangle $HN = AX \cdot XB - BX^2$.

---

*The term "golden section" is also used. See Beman and Smith's *New Plane and Solid Geometry*, p. 196.

**58.** Let $AB = a$, $XB = x$.

Then $(a - x)^2 = ax$, by § 51.

$$a^2 + x^2 = 3ax, \text{ by § 54};$$

$$\therefore x^2 - 3ax + a^2 = 0$$

and $x = \dfrac{a}{2}(3 - \sqrt{5})$.

$$\therefore x^2 = \frac{a^2}{2}(7 - 3\sqrt{5}).$$

$$\therefore a - x = \frac{a}{2}(\sqrt{5} - 1) = a \times 0.6180\ldots.$$

$$\therefore (a - x)^2 = \frac{a^2}{2}(3 - \sqrt{5}) = a^2 \times 0.3819\ldots.$$

The rect. $BPKX$

$$= (a - x)x$$

$$= a^2(\sqrt{5} - 2) = a^2 \times 0.2360\ldots.$$

$$EA^2 = 5EB^2 = \frac{5}{4}AB^2.$$

$$EA = \frac{\sqrt{5}}{2}AB = 1.1180\ldots \times a.$$

**59.** In the language of proportion

$$AB : AX = AX : XB.$$

The straight line $AB$ is said to be divided "in extreme and mean ratio."

**60.** Let $AB$ be divided in $X$ in median section. Complete the rectangle $CBXH$ (Fig. 19). Bisect the rectangle by the line $MNO$. Find the point $N$ by laying $XA$ over $X$ so that $A$ falls on $MO$, and fold through $XN$, $NB$, and $NA$. Then $BAN$ is an isosceles triangle having its angles $ABN$ and $BNA$ double the angle $NAB$.

$$AX = XN = NB$$
$$\angle ABN = \angle NXB$$
$$\angle NAX = \angle XNA$$
$$\angle NXB = 2\angle NAX$$

Fig. 19.

$$\angle ABN = 2\angle NAB.$$
$$AN^2 = MN^2 + AM^2$$
$$= BN^2 - BM^2 + AM^2$$
$$= AX^2 + AB \cdot AX$$
$$= AB \cdot XB + AB \cdot AX$$
$$= AB^2$$

$$\therefore AN = AB$$

and $\qquad \angle NAB = \frac{2}{5}$ of a right angle.

**61.** The right angle at $A$ can be divided into five equal parts as in Fig. 20. Here $N'$ is found as in § 60. Then fold $AN'Q$; bisect $\angle QAB$ by folding,

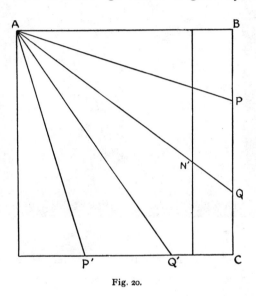

Fig. 20.

fold over the diagonal $AC$ and thus get the point $Q', P'$.

**62.** To describe a right-angled triangle, given the hypotenuse $AB$, and the altitude.

Fold $EF$ (Fig. 21) parallel to $AB$ at the distance of the given altitude.

Take $G$ the middle point of $AB$. Find $H$ by folding $GB$ through $G$ so that $B$ may fall on $EF$.

Fold through *H* and *A*, *G*, and *B*.

*AHB* is the triangle required.

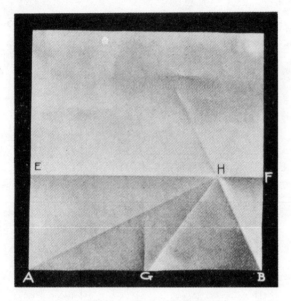

Fig. 21.

**63.** *ABCD* (Fig. 22) is a rectangle. It is required to find a square equal to it in area.

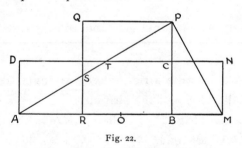

Fig. 22.

Mark off *BM* = *BC*.

Find *O*, the middle point of *AM*, by folding.

Fold *OM*, keeping *O* fixed and letting *M* fall on line *BC*, thus finding *P*, the vertex of the right-angled triangle *AMP*.

Describe on *PB* the square *BPQR*.

The square is equal to the given rectangle.

For ∵ *BP=QP*, and the angles are equal, triangle *BMP* is evidently congruent to triangle *QSP*.

∴ *QS=BM=AD*.

∴ triangles *DAT* and *QSP* are congruent.

∴ *PC=SR* and triangles *RSA* and *CPT* are congruent.

∴ ▭ *ABCD* can be cut into three parts which can be fitted together to form the square *RBPQ*.

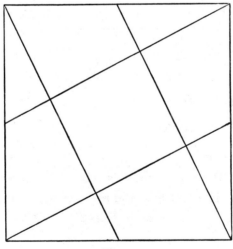

Fig. 23.

**64.** Take four equal squares and cut each of them into two pieces through the middle point of one of the sides and an opposite corner. Take also another equal square. The eight pieces can be arranged round the square so as to form a complete square, as in Fig. 23, the arrangement being a very interesting puzzle.

The fifth square may evidently be cut like the others, thus complicating the puzzle.

**65.** Similar puzzles can be made by cutting the squares through one corner and the trisection points of the opposite side, as in Fig. 24.

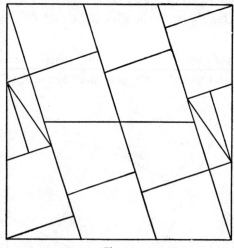

Fig. 24.

**66.** If the nearer point is taken 10 squares are required, as in Fig. 24; if the remoter point is taken 13 squares are required, as in Fig. 25.

**67.** The puzzles mentioned in §§ 65, 66, are based upon the formulas

$$1^2 + 2^2 = 5$$
$$1^2 + 3^2 = 10$$
$$2^2 + 3^2 = 13.$$

The process may be continued, but the number of squares will become inconveniently large.

**68.** Consider again Fig. 13 in § 44. If the four triangles at the corners of the given square are removed, one square is left. If the two rectangles *FK* and *KG* are removed, two squares in juxtaposition are left.

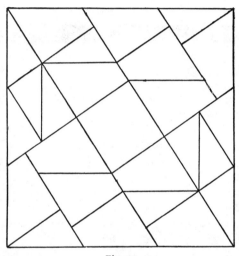

Fig. 25.

**69.** The given square may be cut into pieces which can be arranged into two squares. There are various ways of doing this. Fig. 23, in § 65, suggests the following elegant method: The required pieces are (1) the square in the center, and (2) the four congruent symmetric quadrilaterals at the corners, together with the four triangles. In this figure the lines from the mid-points of the sides pass through the corners of the given square, and the central square is one fifth of it. The magnitude of the inner square can be varied by taking other points on the sides instead of the corners.

**70.** The given square can be divided as follows (Fig. 26) into three equal squares:

Take $BG =$ half the diagonal of the square.

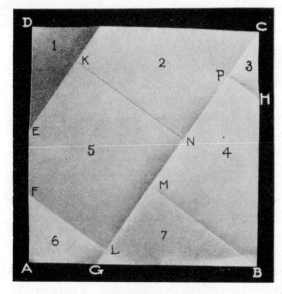

Fig 26.

Fold through $C$ and $G$.

Fold $BM$ perpendicular to $CG$.

Take $MP$, $CN$, and $NL$ each $= BM$.

Fold $PH$, $NK$, $LF$ at right angles to $CG$, as in Fig. 26.

Take $NK = BM$, and fold $KE$ at right angles to $NK$.

Then the pieces 1, 4, and 6, 3 and 5, and 2 and 7 form three equal squares.

Now $CG^2 = 3BG^2$,

and from the triangles $GBC$ and $CMB$

$$\frac{BM}{BC} = \frac{BG}{CG};$$

Letting $BC = a$, we have

$$BM = \frac{a}{\sqrt{3}}.$$

# IV. THE PENTAGON.

**71.** To cut off a regular pentagon from the square *ABCD*.

Divide *BA* in *X* in median section and take *M* the mid-point of *AX*.

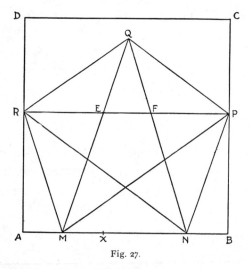

Fig. 27.

Then $AB \cdot AX = XB^2$, and $AM = MX$.

Take $BN = AM$ or $MX$.

Then $MN = XB$.

Lay off *NP* and *MR* equal to *MN*, so that *P* and *R* may lie on *BC* and *AD* respectively.

Lay off $RQ$ and $PQ = MR$ and $NP$.

$MNPQR$ is the pentagon required.

In Fig. 19, p. 22, $AN$, which is equal to $AB$, has the point $N$ on the perpendicular $MO$. If $A$ be moved on $AB$ over the distance $MB$, then it is evident that $N$ will be moved on to $BC$, and $X$ to $M$.

Therefore, in Fig. 27, $NR = AB$. Similarly $MP = AB$. $RP$ is also equal to $AB$ and parallel to it.

$$\angle RMA = \tfrac{4}{5} \text{ of a rt. } \angle .$$

$$\therefore \angle NMR = \tfrac{6}{5} \text{ of a rt. } \angle .$$

Similarly $\quad \angle PNM = \tfrac{6}{5}$ of a rt. $\angle$ .

From triangles $MNR$ and $QRP$, $\angle NMR = \angle RQP = \tfrac{6}{5}$ of a rt. $\angle$ .

The three angles at $M$, $N$, and $Q$ of the pentagon being each equal to $\tfrac{6}{5}$ of a right angle, the remaining two angles are together equal to $\tfrac{12}{5}$ of a right angle, and they are equal. Therefore each of them is $\tfrac{6}{5}$ of a right angle.

Therefore all the angles of the pentagon are equal.

The pentagon is also equilateral by construction.

**72.** The base $MN$ of the pentagon is equal to $XB$, i. e., to $\dfrac{AB}{2} \cdot (\sqrt{5} - 1) = AB \times 0.6180\ldots$ § 58.

The greatest breadth of the pentagon is $AB$.

**73.** If $p$ be the altitude,

$$AB^2 = p^2 + \left[ \frac{AB}{4} (\sqrt{5} - 1) \right]^2$$

$$= p^2 + AB^2 \cdot \frac{3 - \sqrt{5}}{8}.$$

$$\therefore p^2 = AB^2 \cdot \left(1 - \frac{3 - \sqrt{5}}{8}\right)$$

$$= AB^2 \cdot \frac{5 + \sqrt{5}}{8}.$$

$$\therefore p = AB \cdot \frac{\sqrt{10 + 2\sqrt{5}}}{4}$$

$$= AB \times 0.9510\ldots = AB \cos 18°.$$

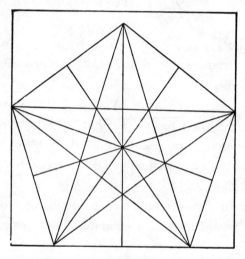

Fig. 28.

**74.** If $R$ be the radius of the circumscribed circle,

$$R = \frac{AB}{2 \cos 18°} = \frac{2AB}{\sqrt{10 + 2\sqrt{5}}}$$

$$= AB \cdot \sqrt{\frac{5 - \sqrt{5}}{10}}$$

$$= AB \times 0.5257\ldots$$

**75.** If $r$ be the radius of the inscribed circle, then from Fig. 28 it is evident that

$$r = p - R = AB \cdot \sqrt{\frac{5 + \sqrt{5}}{8}} - AB \cdot \sqrt{\frac{5 - \sqrt{5}}{10}}$$

$$= AB \cdot \sqrt{5 + \sqrt{5}} \left( \sqrt{\frac{1}{8}} - \sqrt{\frac{3 - \sqrt{5}}{20}} \right)$$

$$= AB \cdot \sqrt{5 + \sqrt{5}} \left[ \frac{\sqrt{5} - (\sqrt{5} - 1)}{\sqrt{40}} \right]$$

$$= AB \cdot \sqrt{\frac{5 + \sqrt{5}}{40}}$$

$$= AB \times 0.4253 \ldots$$

**76.** The area of the pentagon is $5r \times \frac{1}{2}$ the base of the pentagon, i. e.,

$$5 AB \cdot \sqrt{\frac{5 + \sqrt{5}}{40}} \cdot \frac{AB}{4} \cdot (\sqrt{5} - 1)$$

$$= AB^2 \cdot \frac{5}{4} \cdot \sqrt{\frac{5 - \sqrt{5}}{10}} = AB^2 \times 0.6571 \ldots$$

**77.** In Fig. 27 let $PR$ be divided by $MQ$ and $NQ$ in $E$ and $F$.

Then $\because MN = \dfrac{AB}{2} \cdot (\sqrt{5} - 1) \ldots$ § 72

and $\cos 36° = \dfrac{\frac{1}{2}AB}{\frac{1}{2}AB \cdot (\sqrt{5} - 1)}$,

$\therefore RE = FP = \dfrac{MN}{2} \cdot \dfrac{1}{\cos 36°} = AB \cdot \dfrac{\sqrt{5} - 1}{\sqrt{5} + 1}$

$$= AB \cdot \frac{3 - \sqrt{5}}{2} \ldots (1)$$

$$EF = AB - 2RE = AB - AB(3 - \sqrt{5})$$
$$= AB(\sqrt{5} - 2) \dots (2)$$

$RF = MN.$

$$RF : RE = RE : EF \text{ (by § 51)} \dots\dots\dots\dots\dots (3)$$
$$\sqrt{5} - 1 : 3 - \sqrt{5} = 3 - \sqrt{5} : 2(\sqrt{5} - 2) \dots\dots\dots (4)$$

By § 76 the area of the pentagon

$$= AB^2 \cdot \frac{5}{4} \cdot \sqrt{\frac{5 - \sqrt{5}}{10}}$$

$$= MN^2 \cdot \left(\frac{\sqrt{5} + 1}{2,}\right)^2 \cdot \frac{5}{4} \cdot \sqrt{\frac{5 - \sqrt{5}}{10}}$$

$$= MN^2 \cdot \frac{1}{4} \cdot \sqrt{25 + 10\sqrt{5}},$$

since $AB = MN \cdot \dfrac{\sqrt{5} + 1}{2}.$

$\therefore$ the area of the inner pentagon

$$= EF^2 \cdot \frac{1}{4} \cdot \sqrt{25 + 10\sqrt{5}}$$

$$= AB^2 \cdot (\sqrt{5} - 2)^2 \cdot \frac{1}{4} \cdot \sqrt{25 + 10\sqrt{5}}.$$

The larger pentagon divided by the smaller

$$= MN^2 : EF^2$$
$$= 2 : (7 - 3\sqrt{5})$$
$$= 1 : 0.145898\dots.$$

**78.** If in Fig. 27, angles $QEK$ and $LFQ$ are made equal to $ERQ$ or $FQP$, $K$, $L$ being points on the sides $QR$ and $QP$ respectively, then $EFLQK$ will be a regular pentagon congruent to the inner pentagon. Pentagons can be similarly described on the remaining sides of the inner pentagon. The resulting figure consisting of six pentagons is very interesting.

# V. THE HEXAGON.

**79.** To cut off a regular hexagon from a given square.

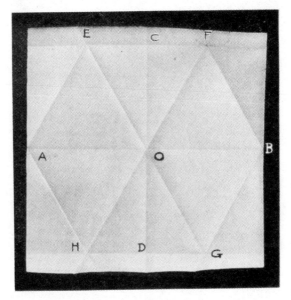

Fig. 29.

Fold through the mid-points of the opposite sides, and obtain the lines $AOB$ and $COD$.

On both sides of $AO$ and $OB$ describe equilateral triangles (§ 25), $AOE$, $AHO$; $BFO$ and $BOG$.

Draw *EF* and *HG*.

*AHGBFE* is a regular hexagon.

It is unnecessary to give the proof.

The greatest breadth of the hexagon is *AB*.

**80.** The altitude of the hexagon is

$$\frac{\sqrt{3}}{2} \cdot AB = 0.866\ldots \times AB.$$

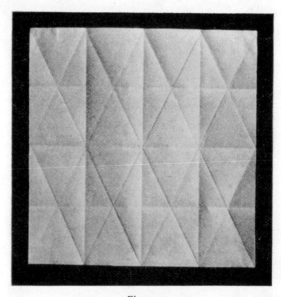

Fig. 30.

**81.** If *R* be the radius of the circumscribed circle,

$$R = \frac{1}{2} AB.$$

**82.** If *r* be the radius of the inscribed circle,

$$r = \frac{\sqrt{3}}{4} \cdot AB = 0.433\ldots \times AB.$$

**83.** The area of the hexagon is 6 times the area of the triangle $HGO$,

$$= 6 \cdot \frac{AB}{4} \cdot \frac{\sqrt{3}}{4} AB.$$

$$= \frac{3\sqrt{3}}{8} \cdot AB^2 = 0.6495\ldots \times AB^2.$$

Also the hexagon $= \frac{3}{4} \cdot AB \cdot CD.$

$= 1\frac{1}{2}$ times the equilateral triangle on $AB$.

Fig. 31.

**84.** Fig. 30 is an example of ornamental folding into equilateral triangles and hexagons.

**85.** A hexagon is formed from an equilateral triangle by folding the three corners to the center.

The side of the hexagon is $\frac{1}{3}$ of the side of the equilateral triangle.

The area of the hexagon $= \frac{2}{3}$ of the equilateral triangle.

**86.** The hexagon can be divided into equal regular hexagons and equilateral triangles as in Fig. 31 by folding through the points of trisection of the sides.

# VI. THE OCTAGON.

**87.** To cut off a regular octagon from a given square.

Obtain the inscribed square by joining the mid-points $A$, $B$, $C$, $D$ of the sides of the given square.

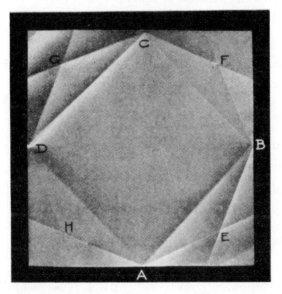

Fig. 32.

Bisect the angles which the sides of the inscribed square make with the sides of the other. Let the bisecting lines meet in $E$, $F$, $G$, and $H$.

*AEBFCGDH* is a regular octagon.

The triangles *AEB*, *BFC*, *CGD*, and *DHA* are congruent isosceles triangles. The octagon is therefore equilateral.

The angles at the vertices, *E*, *F*, *G*, *H* of the same four triangles are each one right angle and a half, since the angles at the base are each one-fourth of a right angle.

Therefore the angles of the octagon at *A*, *B*, *C*, and *D* are each one right angle and a half.

Thus the octagon is equiangular.

The greatest breadth of the octagon is the side of the given square, *a*.

**88.** If *R* be the radius of the circumscribed circle, and *a* be the side of the original square,

$$R = \frac{a}{2}.$$

**89.** The angle subtended at the center by each of the sides is half a right angle.

**90.** Draw the radius *OE* and let it cut *AB* in *K* (Fig. 33).

Then $AK = OK = \dfrac{OA}{\sqrt{2}} = \dfrac{a}{2\sqrt{2}}.$

$$KE = OA - OK = \frac{a}{2} - \frac{a}{2\sqrt{2}} = \frac{a}{4} \cdot (2 - \sqrt{2}).$$

Now from triangle *AEK*,

$$AE^2 = AK^2 + KE^2$$

$$= \frac{a^2}{8} + \frac{a^2}{8} \cdot (3 - 2\sqrt{2})$$

$$= \frac{a^2}{8} \cdot (4 - 2\sqrt{2})$$

$$= \frac{a^2}{4} \cdot (2 - \sqrt{2}).$$

$$\therefore AE = \frac{a}{2} \cdot \sqrt{2 - \sqrt{2}}.$$

**91.** The altitude of the octagon is $CE$ (Fig. 33). But $CE^2 = AC^2 - AE^2$

$$= a^2 - \frac{a^2}{4} \cdot (2 - \sqrt{2}) = \frac{a^2}{4} \cdot (2 + \sqrt{2}).$$

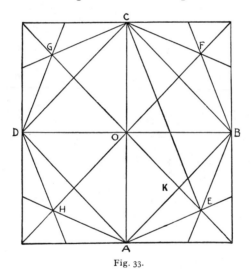

Fig. 33.

$$\therefore CE = \frac{a}{2} \cdot \sqrt{2 + \sqrt{2}}.$$

**92.** The area of the octagon is eight times the triangle $AOE$ and

$$= 4OE \cdot AK = 4 \cdot \frac{a}{2} \cdot \frac{a}{2\sqrt{2}} = \frac{a^2}{\sqrt{2}}.$$

**93.** A regular octagon may also be obtained by dividing the angles of the given square into four equal parts.

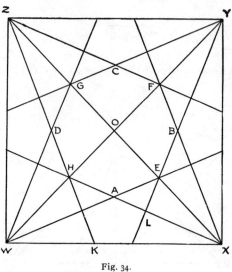

Fig. 34.

It is easily seen that $EZ = WZ = a$, the side of the square.

$$XZ = a\sqrt{2};$$
$$XE = a(\sqrt{2} - 1);$$
$$XE = WH = WK;$$
$$KX = a - a(\sqrt{2} - 1)$$
$$= a(2 - \sqrt{2}).$$

Now $KZ^2 = a^2 + a^2(\sqrt{2} - 1)^2 = a^2(4 - 2\sqrt{2})$

$\therefore KZ = a\sqrt{4 - 2\sqrt{2}}$.

Also $GE = XZ - 2XE$

$$= a\sqrt{2} - 2a(\sqrt{2} - 1)$$
$$= a(2 - \sqrt{2}).$$

$$\therefore HO = \frac{a}{2}(2 - \sqrt{2}).$$

Again $OZ = \frac{a}{2}\sqrt{2}$,

and $HZ^2 = HO^2 + OZ^2$

$$= \frac{a^2}{4}(6 - 4\sqrt{2} + 2)$$

$$= a^2(2 - \sqrt{2}).$$

$$\therefore HZ = a\sqrt{2 - \sqrt{2}}.$$

$$HK = KZ - HZ$$

$$= a\sqrt{4 - 2\sqrt{2}} - a\sqrt{2 - \sqrt{2}}$$

$$= a \cdot (\sqrt{2 - \sqrt{2}}) \cdot (\sqrt{2} - 1)$$

$$= a\sqrt{10 - 7\sqrt{2}}.$$

$$AL = \frac{1}{2}HK = \frac{a}{2}\sqrt{10 - 7\sqrt{2}},$$

and $HA = \frac{a}{2}\sqrt{20 - 14\sqrt{2}}.$

**94.** The area of the octagon is eight times the area of the triangle *HOA*,

$$= 8 \cdot \tfrac{1}{2}HO \cdot \frac{HO}{\sqrt{2}}$$

$$= HO^2 \cdot 2\sqrt{2}$$

$$= \left[\frac{a}{2}(2 - \sqrt{2})\right]^2 \cdot 2\sqrt{2}$$

$$= \frac{a^2}{4} \cdot 2\sqrt{2} \cdot (6 - 4\sqrt{2})$$

$$= a^2 \cdot (3\sqrt{2} - 4)$$

$$= a^2 \cdot \sqrt{2} \cdot (\sqrt{2} - 1)^2.$$

**95.** This octagon : the octagon in § 92

$$= (2 - \sqrt{2})^2 : 1 \text{ or } 2 : (\sqrt{2} + 1)^2;$$

and their bases are to one another as

$$\sqrt{2} : \sqrt{2} + 1.$$

# VII. THE NONAGON.

**96.** Any angle can be trisected fairly accurately by paper folding, and in this way we may construct approximately the regular nonagon.

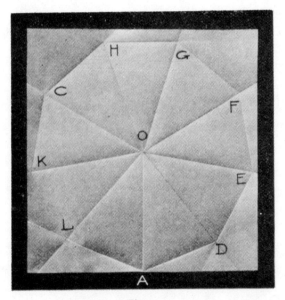

Fig. 35.

Obtain the three equal angles at the center of an equilateral triangle. (§ 25.)

For convenience of folding, cut out the three angles, $AOF$, $FOC$, and $COA$.

Trisect each of the angles as in Fig. 35, and make each of the arms $= OA$.

**97**. Each of the angles of a nonagon is $\frac{14}{9}$ of a right angle $= 140°$.

The angle subtended by each side at the center is $\frac{4}{9}$ of a right angle or $40°$.

Half this angle is $\frac{1}{7}$ of the angle of the nonagon.

**98**. $OA = \frac{1}{2}a$, where $a$ is the side of the square; it is also the radius of the circumscribed circle, $R$.

The radius of the inscribed circle $= R \cdot \cos 20°$

$$= \tfrac{1}{2}a \cos 20°$$

$$= \frac{a}{2} \times 0.9396926$$

$$= a \times 0.4698463.$$

The area of the nonagon is 9 times the area of the triangle $AOL$

$$= 9 \cdot R \cdot \tfrac{1}{2} R \sin 40°$$

$$= \tfrac{9}{2} R^2 \cdot \sin 40°$$

$$= \frac{9a^2}{8} \times 0.6427876$$

$$= a^2 \times 0.723136.$$

# VIII. THE DECAGON AND THE DODECAGON

**99.** Figs. 36, 37 show how a regular decagon, and a regular dodecagon, may be obtained from a penta-gon and hexagon respectively.

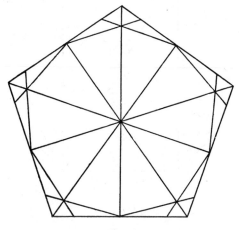

Fig. 36.

The main part of the process is to obtain the angles at the center.

In Fig. 36, the radius of the inscribed circle of the pentagon is taken for the radius of the circum-scribed circle of the decagon, in order to keep it within the square.

**100.** A regular decagon may also be obtained as follows:

Obtain *X, Y,* (Fig. 38), as in § 51, dividing *AB* in median section.

Take *M* the mid-point of *AB*.

Fold *XC, MO, YD* at right angles to *AB*.

Take *O* in *MO* such that *YO = AY,* or *YO = XB*.

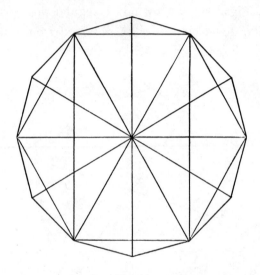

Fig. 37.

Let *YO*, and *XO* produced meet *XC*, and *YD* in *C* and *D* respectively.

Divide the angles *XOC* and *DOY* into four equal parts by *HOE, KOF,* and *LOG*.

Take *OH, OK, OL, OE, OF,* and *OG* equal to *OY* or *OX*.

Join *X, H, K, L, C, D, E, F, G,* and *Y*, in order.

As in § 60,

$$\angle YOX = \tfrac{2}{5} \text{ of a rt. } \angle = 36°.$$

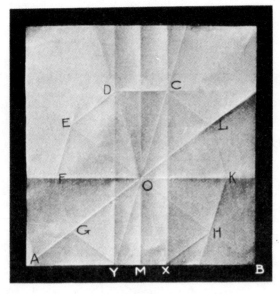

Fig. 38.

By bisecting the sides and joining the points thus determined with the center, the perigon is divided into sixteen equal parts. A 16-gon is therefore easily constructed, and so for a 32-gon, and in general a regular $2^n$-gon.

# IX. THE PENTEDECAGON.

**101.** Fig. 39 shows how the pentedecagon is obtained from the pentagon.

Let *ABCDE* be the pentagon and *O* its center.

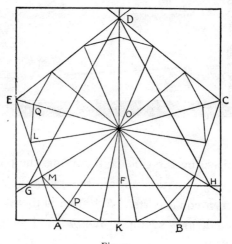

Fig. 39.

Draw *OA*, *OB*, *OC*, *OD*, and *OE*. Produce *DO* to meet *AB* in *K*.

Take $OF = \frac{1}{2}$ of *OD*.

Fold *GFH* at right angles to *OF*. Make *OG* = *OH* = *OD*.

Then $GDH$ is an equilateral triangle, and the angles $DOG$ and $HOD$ are each $120°$.

But angle $DOA$ is $144°$; therefore angle $GOA$ is $24°$.

That is, the angle $EOA$, which is $72°$, is trisected by $OG$.

Bisect the angle $EOG$ by $OL$, meeting $EA$ in $L$, and let $OG$ cut $EA$ in $M$; then

$$OL = OM.$$

In $OA$ and $OE$ take $OP$ and $OQ$ equal to $OL$ or $OM$.

Then $PM$, $ML$, and $LQ$ are three sides of the pentedecagon.

Treating similarly the angles $AOB$, $BOC$, $COD$, and $DOE$, we obtain the remaining sides of the pentedecagon.

# X. SERIES.

**102.** Fig. 40 illustrates an arithmetic series. The horizontal lines to the left of the diagonal, including the upper and lower edges, form an arithmetic series.

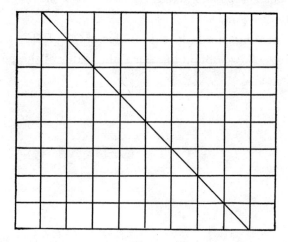

Fig. 40.

The initial line being $a$, and $d$ the common difference, the series is $a$, $a + d$, $a + 2d$, $a + 3d$, etc.

**103.** The portions of the horizontal lines to the right of the diagonal also form an arithmetic series, but they are in reverse order and decrease with a common difference.

**104**. In general, if $l$ be the last term, and $s$ the sum of the series, the above diagram graphically proves the formula

$$s = \frac{n}{2}(a + l).$$

**105.** If $a$ and $c$ are two alternate terms, the middle term is

$$\frac{a + c}{2}.$$

**106.** To insert $n$ means between $a$ and $l$, the vertical line has to be folded into $n + 1$ equal parts. The common difference will be

$$\frac{l - a}{n + 1}.$$

**107.** Considering the reverse series and interchanging $a$ and $l$, the series becomes

$$a, \ a - d, \ a - 2d \ldots . l.$$

The terms will be positive so long as $a > (n-1)d$, and thereafter they will be zero or negative.

### GEOMETRIC SERIES.

**108.** In a right-angled triangle, the perpendicular from the vertex on the hypotenuse is a geometric mean between the segments of the hypotenuse. Hence, if two alternate or consecutive terms of a geometric series are given in length, the series can be determined as in Fig. 41. Here $OP_1$, $OP_2$, $OP_3$, $OP_4$, and $OP_5$ form a geometric series, the common rate being $OP_1 : OP_2$.

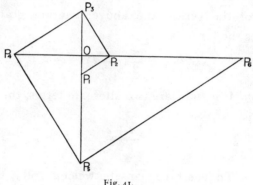

Fig. 41.

If $OP_1$ be the unit of length, the series consists of the natural powers of the common rate.

**109.** Representing the series by $a$, $ar$, $ar^2$, ....

$$P_1 P_2 = a\sqrt{1 + r^2}.$$
$$P_2 P_3 = ar\sqrt{1 + r^2}.$$
$$P_3 P_4 = ar^2\sqrt{1 + r^2}.$$

. . . . . . . . . . . . . . . . .

These lines also form a geometric series with the common rate $r$.

**110.** The terms can also be reversed, in which case the common rate will be a proper fraction. If $OP_5$ be the unit, $OP_4$ is the common rate. The sum of the series to infinity is

$$\frac{OP_5}{OP_5 - OP_4}.$$

**111.** In the manner described in § 108, one geometric mean can be found between two given lines, and by continuing the process, 3, 7, 15, etc., means can be found. In general, $2^n - 1$ means can be found, $n$ being any positive integer.

**112.** It is not possible to find two geometric means between two given lines, merely by folding through known points. It can, however, be accomplished in the following manner : In Fig. 41, $OP_1$ and $OP_4$ being given, it is required to find $P_2$ and $P_3$. Take two rectangular pieces of paper and so arrange them, that their outer edges pass through $P_1$ and $P_4$, and two corners lie on the straight lines $OP_2$ and $OP_3$ in such a way that the other edges ending in those corners coincide. The positions of the corners determine $OP_2$ and $OP_3$.

**113.** This process gives the cube root of a given number, for if $OP_1$ is the unit, the series is 1, $r$, $r^2$, $r^3$.

**114.** There is a very interesting legend in connection with this problem.* "The Athenians when suffering from the great plague of eruptive typhoid fever in 430 B. C., consulted the oracle at Delos as to how they could stop it. Apollo replied that they must *double* the *size* of his altar which was in the form of a *cube*. Nothing seemed more easy, and a new altar was constructed having each of its *edges* double that of the old one. The god, not unnaturally indignant,

---

*But see Beman and Smith's translation of Fink's *History of Mathematics*, p. 82, 207.

made the pestilence worse than before. A fresh deputation was accordingly sent to Delos, whom he informed that it was useless to trifle with him, as he must have his altar exactly doubled. Suspecting a mystery, they applied to the geometricians. Plato, the most illustrious of them, declined the task, but referred them to Euclid, who had made a special study of the problem." (Euclid's name is an interpolation for that of Hippocrates.) Hippocrates reduced the question to that of finding two geometric means between two straight lines, one of which is twice as long as the other. If $a$, $x$, $y$ and $2a$ be the terms of the series, $x^3 = 2a^3$. He did not, however, succeed in finding the means. Menaechmus, a pupil of Plato, who lived between 375 and 325 B. C., gave the following three equations : *

$$a : x = x : y = y : 2a.$$

From this relation we obtain the following three equations :

$$x^2 = ay \dots\dots\dots\dots\dots (1)$$
$$y^2 = 2ax \dots\dots\dots\dots\dots (2)$$
$$xy = 2a^2 \dots\dots\dots\dots\dots (3)$$

(1) and (2) are equations of parabolas and (3) is the equation of a rectangular hyperbola. Equations (1) and (2) as well as (1) and (3) give $x^3 = 2a^3$. The problem was solved by taking the intersection ($\alpha$) of the two parabolas (1) and (2), and the intersection ($\beta$) of the parabola (1) with the rectangular hyperbola (3).

*Ibid., p. 207.

### HARMONIC SERIES.

**115.** Fold any lines *AR*, *PB*, as in Fig. 42, *P* being on *AR*, and *B* on the edge of the paper. Fold again so that *AP* and *PR* may both coincide with *PB*. Let *PX*, *PY* be the creases thus obtained, *X* and *Y* being on *AB*.

Then the points *A*, *X*, *B*, *Y* form an harmonic range. That is, *AB* is divided internally in *X* and externally in *Y* so that

$$AX : XB = AY : BY.$$

I⸱ is evident, that every line cutting *PA*, *PX*, *PB*, and *PY* will be divided harmonically.

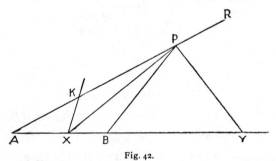

Fig. 42.

**116.** Having given *A*, *B*, and *X*, to find *Y*: fold any line *XP* and mark *K* corresponding to *B*. Fold *AKPR*, and *BP*. Bisect the angle *BPR* by *PY* by folding through *P* so that *PB* and *PR* coincide.

Because *XP* bisects the angle *APB*,

$$\therefore AX : XB = AP : BP,$$
$$= AY : BY.$$

**117.**
$$AX : XB = AY : BY$$
$$\text{or } AY - XY : XY - BY = AY : BY.$$

Thus, $AY$, $XY$, and $BY$, are an harmonic series, and $XY$ is the harmonic mean between $AY$ and $BY$.

Similarly $AB$ is the harmonic mean between $AX$ and $AY$.

**118.** If $BY$ and $XY$ be given, to find the third term $AY$, we have only to describe any right-angled triangle on $XY$ as the hypotenuse and make angle $APX$ = angle $XPB$.

**119.** Let $AX = a$, $AB = b$, and $AY = c$.

$$\text{Then } b = \frac{2ac}{a + c};$$
$$\text{or, } ab + bc = 2ac$$
$$\text{or, } c = \frac{ab}{2a - b} = \frac{b}{2 - \dfrac{b}{a}}.$$

When $a = b$, $c = b$.

When $b = 2a$, $c = \infty$.

Therefore when $X$ is the middle point of $AB$, $Y$ is at an infinite distance to the right of $B$. $Y$ approaches $B$ as $X$ approaches it, and ultimately the three points coincide.

As $X$ moves from the middle of $AB$ to the left, $Y$ moves from an infinite distance on the left towards $A$, and ultimately $X$, $A$, and $Y$ coincide.

**120.** If $E$ be the middle point of $AB$,
$$EX \cdot EY = EA^2 = EB^2$$
for all positions of $X$ and $Y$ with reference to $A$ or $B$.

Each of the two systems of pairs of points $X$ and $Y$ is called a system in involution, the point $E$ being called the center and $A$ or $B$ the focus of the system. The two systems together may be regarded as one system.

**121.** $AX$ and $AY$ being given, $B$ can be found as follows:

Produce $XA$ and take $AC = XA$.

Take $D$ the middle point of $AY$.

Take $CE = DA$ or $AE = DC$.

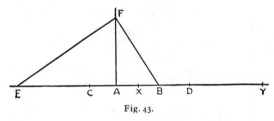

Fig. 43.

Fold through $A$ so that $AF$ may be at right angles to $CAY$.

Find $F$ such that $DF = DC$.

Fold through $EF$ and obtain $FB$, such that $FB$ is at right angles to $EF$.

$CD$ is the arithmetic mean between $AX$ and $AY$.

$AF$ is the geometric mean between $AX$ and $AY$.

$AF$ is also the geometric mean between $CD$ or $AE$ and $AB$.

Therefore $AB$ is the harmonic mean between $AX$ and $AY$.

**122.** The following is a very simple method of finding the harmonic mean between two given lines.

Take *AB*, *CD* on the edges of the square equal to the given lines. Fold the diagonals *AD*, *BC* and the sides *AC*, *BD* of the trapezoid *ACDB*. Fold through *E*, the point of intersection of the diagonals, so that *FEG* may be at right angles to the other sides of the square or parallel to *AB* and *CD*. Let *FEG* cut *AC*

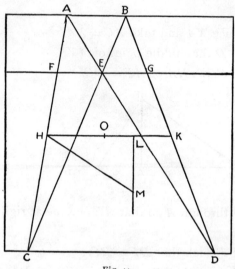

Fig. 44.

and *BD* in *F* and *G*. Then *FG* is the harmonic mean between *AB* and *CD*.

For
$$\frac{FE}{AB} = \frac{CE}{CB}$$

and
$$\frac{EG}{CD} = \frac{FE}{CD} = \frac{EB}{CB}$$

$$\therefore \frac{FE}{AB} + \frac{EF}{CD} = \frac{CE}{CB} + \frac{EB}{CB} = 1.$$

$$\therefore \frac{1}{AB} + \frac{1}{CD} = \frac{1}{FE} = \frac{2}{FG}.$$

**123.** The line $HK$ connecting the mid-points of $AC$ and $BD$ is the arithmetic mean between $AB$ and $CD$.

**124.** To find the geometric mean, take $HL$ in $HK$ $=FG$. Fold $LM$ at right angles to $HK$. Take $O$ the mid-point of $HK$ and find $M$ in $LM$ so that $OM=OH$. $HM$ is the geometric mean between $AB$ and $CD$ as well as between $FG$ and $HK$. The geometric mean between two quantities is thus seen to be the geometric mean between their arithmetic mean and harmonic mean.

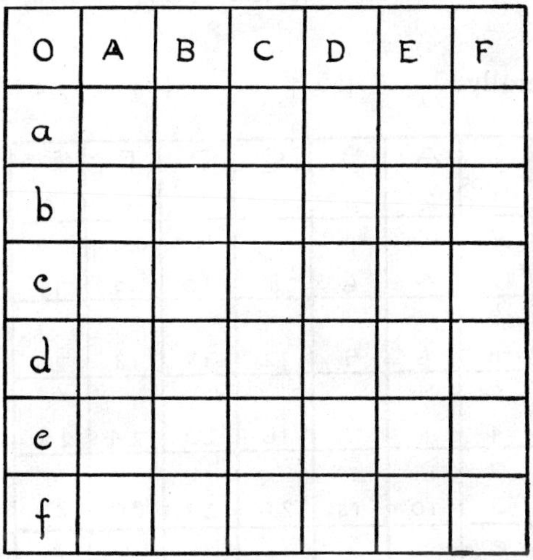

| O | A | B | C | D | E | F |
|---|---|---|---|---|---|---|
| a | | | | | | |
| b | | | | | | |
| c | | | | | | |
| d | | | | | | |
| e | | | | | | |
| f | | | | | | |

Fig. 45.

SUMMATION OF CERTAIN SERIES.

**125.** To sum the series

$$1+3+5\ldots+(2n-1).$$

Divide the given square into a number of equal squares as in Fig. 45. Here we have 49 squares, but the number may be increased as we please.

The number of squares will evidently be a square number, the square of the number of divisions of the sides of the given square.

Let each of the small squares be considered as the unit; the figure formed by $A + O + a$ being called a gnomon.

The numbers of unit squares in each of the gnomons $AOa$, $BOb$, etc., are respectively 3, 5, 7, 9, 11, 13.

Therefore the sum of the series 1, 3, 5, 7, 9, 11, 13 is $7^2$.

Generally, $1 + 3 + 5 + \ldots + (2n - 1) = n^2$.

| O | A | B | C | D | E | F |
|---|---|---|---|---|---|---|
| 1 | 2 | 3 | 4 | 5 | 6 | 7 |
| a | | | | | | |
| 2 | 4 | 6 | 8 | 10 | 12 | 14 |
| b | | | | | | |
| 3 | 6 | 9 | 12 | 15 | 18 | 21 |
| c | | | | | | |
| 4 | 8 | 12 | 16 | 20 | 24 | 28 |
| d | | | | | | |
| 5 | 10 | 15 | 20 | 25 | 30 | 35 |
| e | | | | | | |
| 6 | 12 | 18 | 24 | 30 | 36 | 42 |
| f | | | | | | |
| 7 | 14 | 21 | 28 | 35 | 42 | 49 |

Fig. 46.

**126.** To find the sum of the cubes of the first $n$ natural numbers.

Fold the square into 49 equal squares as in the

preceding article, and letter the gnomons. Fill up the squares with numbers as in the multiplication table.

The number in the initial square is $1 = 1^3$.

The sums of the numbers in the gnomons $Aa$, $Bb$, etc., are $2 + 4 + 2 = 2^3$, $3^3$, $4^3$, $5^3$, $6^3$, and $7^3$.

The sum of the numbers in the first horizontal row is the sum of the first seven natural numbers. Let us call it $s$.

Then the sums of the numbers in rows $a$, $b$, $c$, $d$, etc., are

$$2s, \ 3s, \ 4s, \ 5s, \ 6s, \text{ and } 7s.$$

Therefore the sum of all the numbers is

$$s(1 + 2 + 3 + 4 + 5 + 6 + 7) = s^2.$$

Therefore, the sum of the cubes of the first seven natural numbers is equal to the square of the sum of those numbers.

Generally, $1^3 + 2^3 + 3^3 \ldots + n^3$

$$= (1 + 2 + 3 \ldots + n)^2.$$

$$\therefore \ \Sigma n^3 = \left[ \frac{n(n+1)}{2} \right]^2.$$

For $\quad [n \cdot (n + 1)]^2 - [(n - 1) \cdot n]^2$

$$= (n^2 + n)^2 - (n^2 - n)^2 = 4n^3.$$

Putting $n = 1, 2, 3 \ldots$ in order, we have

$$4 \cdot 1^3 = (1 \cdot 2)^2 - (0 \cdot 1)^2$$
$$4 \cdot 2^3 = (2 \cdot 3)^2 - (1 \cdot 2)^2$$
$$4 \cdot 3^3 = (3 \cdot 4)^2 - (2 \cdot 3)^2$$
$$\ldots = \ldots \ldots$$
$$\ldots = \ldots \ldots$$
$$4 \cdot n^3 = [n \cdot (n + 1)]^2 - [(n - 1) \cdot n]^2.$$

Adding we have

$$4\Sigma n^3 = [n(n+1)]^2$$

$$\therefore \Sigma n^3 = \left[\frac{n(n+1)}{2}\right]^2.$$

**127.** If $s_n$ be the sum of the first $n$ natural numbers,

$$s_n{}^2 - s_n{}^2{}_{-1} = n^3.$$

**128.** To sum the series

$$1 \cdot 2 + 2 \cdot 3 + 3 \cdot 4 \ldots + (n-1) \cdot n.$$

In Fig. 46, the numbers in the diagonal commencing from 1, are the squares of the natural numbers in order.

The numbers in one gnomon can be subtracted from the corresponding numbers in the succeeding gnomon. By this process we obtain

$$n^3 - (n-1)^3 = n^2 - (n-1)^2$$
$$+ 2[n(n-1) + (n-2) + (n-3) \ldots + 1]$$
$$= n^2 + (n-1)^2 + 2[1 + 2 \ldots + (n-1)]$$
$$= n^2 + (n-1)^2 + n(n-1)$$
$$= [n-(n-1)]^2 + 3(n-1)n$$
$$= 1 + 3(n-1)n.$$

Now $\because n^3 - (n-1)^3 = 1 + 3(n-1)n,$

$$\therefore (n-1)^3 - (n-2)^3 = 1 + 3(n-2)(n-1)$$

$$\ldots\ldots\ldots\ldots\ldots\ldots\ldots\ldots$$

$$2^3 - 1^3 = 1 + 3 \cdot 2 \cdot 1$$
$$1^3 - 0^3 = 1 + 0.$$

Hence, by addition,

$$n^3 = n + 3[1 \cdot 2 + 2 \cdot 3 + \ldots + (n-1) \cdot n].$$

Therefore

$$1 \cdot 2 + 2 \cdot 3 \dots + (n-1) \cdot n = \frac{n^3 - n}{3} = \frac{(n-1)n(n+1)}{3}.$$

**129.** To find the sum of the squares of the first $n$ natural numbers.

$$1 \cdot 2 + 2 \cdot 3 \dots + (n-1) \cdot n$$
$$= 2^2 - 2 + 3^2 - 3 \dots + n^2 - n$$
$$= 1^2 + 2^2 + 3^2 \dots + n^2 - (1 + 2 + 3 \dots + n)$$
$$= 1^2 + 2^2 + 3^2 \dots + n^2 - \frac{n(n+1)}{2}.$$

Therefore

$$1^2 + 2^2 + 3^2 \dots + n^2 = \frac{(n-1)n(n+1)}{3} + \frac{n(n+1)}{2}$$
$$= n(n+1)\left[\frac{n-1}{3} + \frac{1}{2}\right]$$
$$= \frac{n(n+1)(2n+1)}{6}.$$

**130.** To sum the series

$$1^2 + 3^2 + 5^2 \dots + (2n-1)^2.$$
$$\therefore n^3 - (n-1)^3 = n^2 + (n-1)^2 + n(n-1), \text{ by § 128,}$$
$$= (2n-1)^2 - (n-1) \cdot n,$$
$$\therefore \text{ by putting } n = 1, 2, 3, \dots$$
$$1^3 - 0^3 = 1^2 - 0 \cdot 1$$
$$2^3 - 1^3 = 3^2 - 1 \cdot 2$$
$$3^3 - 2^3 = 5^2 - 2 \cdot 3$$
$$\dots \dots \dots \dots$$
$$\dots \dots \dots \dots$$
$$n^3 - (n-1)^3 = (2n-1)^2 - (n-1) \cdot n.$$

Adding, we have

$$n^3 = 1^2 + 3^2 + 5^2 \ldots + (2n-1)^2$$
$$- [1 \cdot 2 + 2 \cdot 3 + 3 \cdot 4 \ldots + (n-1) \cdot n].$$
$$\therefore \ 1^2 + 3^2 + 5^2 \ldots + (2n-1)^2$$
$$= n^3 + \frac{n^3 - n}{3}$$
$$= \frac{4n^3 - n}{3} = \frac{n(2n-1)(2n+1)}{3}.$$

# XI. POLYGONS.

**131.** Find $O$ the center of a square by folding its diameters. Bisect the right angles at the center, then the half right angles, and so on. Then we obtain $2^n$ equal angles around the center, and the magnitude of each of the angles is $\dfrac{4}{2^n}$ of a right angle, $n$ being a positive integer. Mark off equal lengths on each of the lines which radiate from the center. If the extremities of the radii are joined successively, we get regular polygons of $2^n$ sides.

**132.** Let us find the perimeters and areas of these polygons. In Fig. 47 let $OA$ and $OA_1$ be two radii at right angles to each other. Let the radii $OA_2$, $OA_3$, $OA_4$, etc., divide the right angle $A_1OA$ into 2, 4, 8.... parts. Draw $AA_1$, $AA_2$, $AA_3$.... cutting the radii $OA_2$, $OA_3$, $OA_4$.... at $B_1$, $B_2$, $B_3$.... respectively, at right angles. Then $B_1$, $B_2$, $B_3$.... are the mid-points of the respective chords. Then $AA_1$, $AA_2$, $AA_3$, $AA_4$.... are the sides of the inscribed polygons of $2^2$, $2^3$, $2^4$.... sides respectively, and $OB_1$, $OB_2$.... are the respective apothems.

Let $OA = R$,

$a(2^n)$ represent the *side* of the inscribed polygon

of $2^n$ sides, $b(2^n)$ the corresponding *apothem*, $p(2^n)$ its *perimeter*, and $A(2^n)$ its *area*.

For the *square*,

$$a(2^2) = R\sqrt{2};$$
$$p(2^2) = R \cdot 2^2 \cdot \sqrt{2};$$
$$b(2^2) = \frac{R}{2}\sqrt{2};$$
$$A(2^2) = R^2 \cdot 2.$$

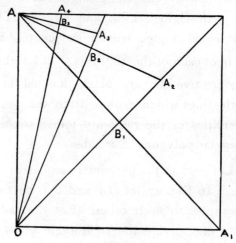

Fig. 47.

For the *octagon*,

in the two triangles $AB_2O$ and $AB_1A_2$

$$\frac{AB_2}{B_1A_2} = \frac{OA}{AA_2}.$$

$$\tfrac{1}{2}AA_2{}^2 = R \cdot B_1A_2 = R[R - b(2^2)]$$

$$= R\left(R - \frac{R}{2}\sqrt{2}\right) = \tfrac{1}{2}R^2 \cdot (2 - \sqrt{2}),$$

or $AA_2 = R\sqrt{2 - \sqrt{2}} = a(2^3)$ ............. (1)

$$p(2^3) = R \cdot 2^3 \sqrt{2 - \sqrt{2}} \quad \ldots \ldots \ldots \ldots \ldots (2)$$

$$b(2^3) = OB_2 = \sqrt{OA^2 - AB_2{}^2} = \sqrt{R^2\left(1 - \frac{2 - \sqrt{2}}{4}\right)}$$

$$= \sqrt{\frac{R^2(2 + \sqrt{2})}{4}} = \tfrac{1}{2}R\sqrt{2 + \sqrt{2}} \quad \ldots (3)$$

$$A(2^3) = \tfrac{1}{2} \text{ perimeter} \times \text{apothem}$$

$$= R \cdot 2^2 \cdot \sqrt{2 - \sqrt{2}} \cdot \tfrac{1}{2}R\sqrt{2 + \sqrt{2}}$$

$$= R^2 \cdot 2\sqrt{2}.$$

Similarly for the polygon of 16 sides,

$$a(2^4) = R\sqrt{2 - \sqrt{2 + \sqrt{2}}};$$

$$p(2^4) = R \cdot 2^4 \cdot \sqrt{2 - \sqrt{2 + \sqrt{2}}};$$

$$b(2^4) = \frac{R}{2}\sqrt{2 + \sqrt{2 + \sqrt{2}}};$$

$$A(2^4) = R^2 \cdot 2^2 \cdot \sqrt{2 - \sqrt{2}};$$

and for the polygon of 32 sides,

$$a(2^5) = R\sqrt{2 - \sqrt{2 + \sqrt{2 + \sqrt{2}}}};$$

$$p(2^5) = R \cdot 2^5 \cdot \sqrt{2 - \sqrt{2 + \sqrt{2 + \sqrt{2}}}};$$

$$b(2^5) = \frac{R}{2}\sqrt{2 + \sqrt{2 + \sqrt{2 + \sqrt{2}}}};$$

$$A(2^5) = R^2 \cdot 2^3 \cdot \sqrt{2 - \sqrt{2 + \sqrt{2}}}.$$

The general law is thus clear.

Also $$A(2^n) = \frac{R}{2} \cdot p(2^{n-1}).$$

As the number of sides is increased indefinitely

the apothem evidently approaches its limit, the radius.  Thus the limit of

$$\sqrt{2 + \sqrt{2 + \sqrt{2}} \ldots} \text{ is } 2 ;$$

for if $x$ represent the limit, $x = \sqrt{2 + x}$, a quadratic which gives $x = 2$, or $-1$; the latter value is, of course, inadmissible.

**133.** If perpendiculars are drawn to the radii at their extremities, we get regular polygons circumscribing the circle and also the polygons described as in the preceding article, and of the same number of sides.

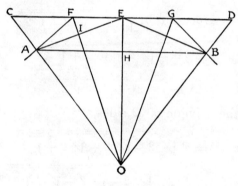

Fig. 48.

In Fig. 48, let $AE$ be a side of the inscribed polygon and $FG$ a side of the circumscribed polygon.

Then from the triangles $FIE$ and $EIO$,

$$\frac{OE}{OI} = \frac{FE}{EI} = \frac{FG}{AE};$$

$$\therefore FG = R \frac{AE}{OI}.$$

The values of $AE$ and $OI$ being known by the previous article, $FG$ is found by substitution.

The areas of the two polygons are to one another as $FG^2 : AE^2$, i. e., as $R^2 : OI^2$.

**134.** In the preceding articles it has been shown how regular polygons can be obtained of $2^2$, $2^3 \ldots 2^n$ sides. And if a polygon of $m$ sides be given, it is easy to obtain polygons of $2^n \cdot m$ sides.

**135.** In Fig. 48, $AB$ and $CD$ are respectively the sides of the inscribed and circumscribed polygons of $n$ sides. Take $E$ the mid point of $CD$ and draw $AE$, $BE$. $AE$ and $BE$ are the sides of the inscribed polygon of $2n$ sides.

Fold $AF$, $BG$ at right angles to $AC$ and $BD$, meeting $CD$ in $F$ and $G$.

Then $FG$ is a side of the circumscribed polygon of $2n$ sides.

Draw $OF$, $OG$ and $OE$.

Let $p$, $P$ be the perimeters of the inscribed and circumscribed polygons respectively of $n$ sides, and $A$, $B$ their areas, and $p'$, $P'$ the perimeters of the inscribed and circumscribed polygons respectively of $2n$ sides, and $A'$, $B'$ their areas.

Then

$$p = n \cdot AB, \; P = n \cdot CD, \; p' = 2n \cdot AE, \; P' = 2n \cdot FG.$$

Because $OF$ bisects $\angle COE$, and $AB$ is parallel to $CD$,

$$\frac{CF}{FE} = \frac{CO}{OE} = \frac{CO}{AO} = \frac{CD}{AB};$$

$$\therefore \frac{CE}{FE} = \frac{CD + AB}{AB},$$

$$\text{or} \ \frac{4n \cdot CE}{4n \cdot FE} = \frac{n \cdot CD + n \cdot AB}{n \cdot AB}$$

$$\therefore \frac{2P}{P'} = \frac{P + p}{p}.$$

$$\therefore P' = \frac{2Pp}{P + p}.$$

Again, from the similar triangles $EIF$ and $AHE$,

$$\frac{EI}{AH} = \frac{EF}{AE},$$

$$\text{or} \ AE^2 = 2AH \cdot EF;$$

$$\therefore \ 4n^2 \cdot AE^2 = 4n^2 \cdot AB \cdot EF,$$

$$\text{or} \ p' = \sqrt{P'p}.$$

Now,

$$A = 2n \triangle AOH, \quad B = 2n \triangle COE,$$

$$A' = 2n \triangle AOE, \quad B' = 4n \triangle FOE.$$

The triangles $AOH$ and $AOE$ are of the same altitude, $AH$,

$$\therefore \frac{\triangle AOH}{\triangle AOE} = \frac{OH}{OE}.$$

Similarly,

$$\frac{\triangle AOE}{\triangle COE} = \frac{OA}{OC}.$$

Again because $AB \parallel CD$,

$$\therefore \frac{\triangle AOH}{\triangle AOE} = \frac{\triangle AOE}{\triangle COE}.$$

$$\therefore \frac{A}{A'} = \frac{A'}{B}, \ \text{or} \ A' = \sqrt{AB}.$$

Now to find $B'$. Because the triangles $COE$ and

*FOE* have the same altitude, and *OF* bisects **the** angle *EOC*,

$$\frac{\triangle COE}{\triangle FOE} = \frac{CE}{FE} = \frac{OC + OE}{OE}.$$

and $OE = OA$,

and $\dfrac{OC}{OA} = \dfrac{OE}{OH} = \dfrac{\triangle AOE}{\triangle AOH};$

$$\therefore \frac{\triangle COE}{\triangle FOE} = \frac{\triangle AOE + \triangle AOH}{\triangle AOH}.$$

From this equation we easily obtain $\dfrac{2B}{B'} = \dfrac{A' + A}{A};$

$$\therefore B' = \frac{2AB}{A + A'}.$$

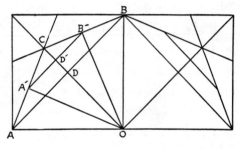

Fig. 49.

**136.** Given the radius $R$ and apothem $r$ of a regular polygon, to find the radius $R'$ and apothem $r'$ of a regular polygon of the *same perimeter* but of *double* the number of sides.

Let $AB$ be a side of the first polygon, $O$ its center, $OA$ the radius of the circumscribed circle, and $OD$ the apothem. On $OD$ produced take $OC = OA$ or $OB$. Draw $AC$, $BC$. Fold $OA'$ and $OB'$ perpendicular to $AC$ and $BC$ respectively, thus fixing the points $A'$, $B'$. Draw $A'B'$ cutting $OC$ in $D'$. Then the chord $A'B'$ is half of $AB$, and the angle $B'OA'$ is half of $BOA$. $OA'$ and $OD'$ are respectively the radius $R'$ and apothem $r'$ of the second polygon.

Now $OD'$ is the arithmetic mean between $OC$ and $OD$, and $OA'$ is the mean proportional between $OC$ and $OD'$.

$$\therefore\ r' = \tfrac{1}{2}(R + r),\ \text{and}\ R' = \sqrt{Rr'}.$$

**137.** Now, take on $OC$, $OE = OA'$ and draw $A'E$. Then $A'D'$ being less than $A'C$, and $\angle D'A'C$ being bisected by $A'E$,

$$ED'\ \text{is less than}\ \tfrac{1}{2}CD',\ \text{i. e., less than}\ \tfrac{1}{4}CD$$
$$\therefore\ R_1 - r_1\ \text{is less than}\ \tfrac{1}{4}(R - r).$$

As the number of sides is increased, the polygon approaches the circle of the same perimeter, and $R$ and $r$ approach the radius of the circle.

That is,

$$R + r + R_1 - r_1 + R_2 - r_2 + \ldots.$$
$$= \text{the diameter of the circle} = \frac{p}{\pi}$$

Also,

$$R_1{}^2 = Rr_1\ \text{or}\ R \cdot \frac{r_1}{R_1} = R_1$$

and $\dfrac{r_2}{R_2} = \dfrac{R_2}{R_1}$, and so on.

Multiplying both sides,

$$R\ \frac{r_1}{R_1} \cdot \frac{r_2}{R_2} \cdot \frac{r_3}{R_3} \ldots. = \text{the radius of the circle} = \frac{p}{2\pi}.$$

138. The radius of the circle lies between $R_n$ and $r_n$, the sides of the polygon being $4 \cdot 2^n$ in number; and $\pi$ lies between $\dfrac{2}{r_n}$ and $\dfrac{2}{R_n}$. The numerical value of $\pi$ can therefore be calculated to any required degree of accuracy by taking a sufficiently large number of sides.

The following are the values of the radii and apothems of the regular polygons of 4, 8, 16....2048 sides.

$$4\text{-gon, } r = 0 \cdot 500000 \quad R = r\sqrt{2} = 0 \cdot 707107$$
$$8\text{-gon, } r_1 = 0 \cdot 603553 \quad R_1 = 0 \cdot 653281$$
$$\cdots\cdots\cdots\cdots\cdots\cdots\cdots\cdots\cdots\cdots\cdots\cdots$$
$$2048\text{-gon, } r_9 = 0 \cdot 636620 \quad R_9 = 0 \cdot 636620.$$
$$\therefore \; \pi = \frac{2}{0 \cdot 636620} = 3 \cdot 14159\ldots$$

139. If $R''$ be the radius of a regular isoperimetric polygon of $4n$ sides

$$R''^2 = \frac{R'^2\,(R + R')}{2R},$$

or in general

$$\frac{R_{k+1}}{R_k} = \sqrt{\frac{1 + \dfrac{R_k}{R_{k-1}}}{2}}.$$

140. The radii $R_1$, $R_2$,....successively diminish, and the ratio $\dfrac{R_2}{R_1}$ is less than unity and equal to the cosine of a certain angle $\alpha$.

$$\frac{R_3}{R_2} = \sqrt{\frac{1 + \cos \alpha}{2}} = \cos \frac{\alpha}{2}.$$

$$\therefore \ \frac{R_{k+1}}{R_k} = \cos \frac{\alpha}{2^{k-1}}$$

multiplying together the different ratios, we get

$$R_{k+1} = R_1 \cdot \cos \alpha \cdot \cos \frac{\alpha}{2} \cdot \cos \frac{\alpha}{2^2} \ldots . \cos \frac{\alpha}{2^{k-1}}$$

The limit of $\cos \alpha \cdot \cos \dfrac{\alpha}{2^2} \ldots . \cos \dfrac{\alpha}{2^{k-1}}$, when $k = \infty$, is $\dfrac{\sin 2\alpha}{2\alpha}$, a result known as *Euler's Formula*.

**141.** It was demonstrated by Karl Friedrich Gauss* (1777–1855) that besides the regular polygons of $2^n$, $3 \cdot 2^n$, $5 \cdot 2^n$, $15 \cdot 2^n$ sides, the only regular polygons which can be constructed by elementary geometry are those the number of whose sides is represented by the product of $2^n$ and one or more different numbers of the form $2^m + 1$. We shall show here how polygons of 5 and 17 sides can be described.

The following theorems are required :†

(1) If $C$ and $D$ are two points on a semi-circumference $ACDB$, and if $C'$ be symmetric to $C$ with respect to the diameter $AB$, and $R$ the radius of the circle,

$$AC \cdot BD = R \cdot (C'D - CD) \ldots \ldots . \text{i.}$$
$$AD \cdot BC = R \cdot (C'D + CD) \ldots \ldots \text{ii.}$$
$$AC \cdot BC = R \cdot CC' \ldots \ldots \ldots \ldots \text{iii.}$$

(2) Let the circumference of a circle be divided into an odd number of equal parts, and let $AO$ be the

---

*Beman and Smith's translation of Fink's *History of Mathematics*, p. 245; see also their translation of Klein's *Famous Problems of Elementary Geometry*, pp. 16, 24, and their *New Plane and Solid Geometry*, p. 212.

†These theorems may be found demonstrated in Catalan's *Théorèmes et Problèmes de Géométrie Elémentaire*.

diameter through one of the points of section $A$ and the mid-point $O$ of the opposite arc. Let the points of section on each side of the diameter be named $A_1$, $A_2$, $A_3$....$A_n$, and $A'_1$, $A'_2$, $A'_3$....$A'_n$ beginning next to $A$.

Then $OA_1 \cdot OA_2 \cdot OA_3 \ldots OA_n = R^n \ldots \ldots$ iv.

and $OA_1 \cdot OA_2 \cdot OA_4 \ldots OA_n = R^{\frac{n}{2}}$.

**142.** It is evident that if the chord $OA_n$ is determined, the angle $A_n OA$ is found and it has only to be divided into $2^n$ equal parts, to obtain the other chords.

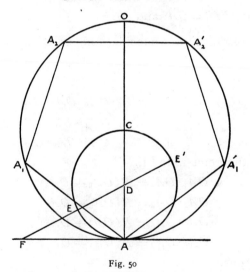

Fig. 50

**143.** Let us first take the pentagon.

By theorem iv,

$$OA_1 \cdot OA_2 = R^2.$$

By theorem i,

$$R(OA_1 - OA_2) = OA_1 \cdot OA_2 = R^2.$$

$$\therefore \ OA_1 - OA_2 = R.$$

$$\therefore \ OA_1 = \frac{R}{2}(\sqrt{5}+1),$$

$$\text{and} \ \ OA_2 = \frac{R}{2}(\sqrt{5}-1).$$

Hence the following construction.

Take the diameter $ACO$, and draw the tangent $AF$. Take $D$ the mid-point of the radius $OC$ and $AF = OC$.

On $OC$ as diameter describe the circle $AE'CE$.

Join $FD$ cutting the inner circle in $E$ and $E'$.

Then $FE' = OA$, and $FE = OA_2$.

**144.** Let us now consider the polygon of seventeen sides.

Here*

$$OA_1 \cdot OA_2 \cdot OA_3 \cdot OA_4 \cdot OA_5 \cdot OA_6 \cdot OA_7 \cdot OA_8 = R^8.$$

$$OA_1 \cdot OA_2 \cdot OA_4 \cdot OA_8 = R^4.$$

$$\text{and} \ \ OA_3 \cdot OA_5 \cdot OA_6 \cdot OA_7 = R^4.$$

By theorems i. and ii.

$$OA_1 \cdot OA_4 = R(OA_3 + OA_5)$$
$$OA_2 \cdot OA_8 = R(OA_6 - OA_7)$$
$$OA_3 \cdot OA_5 = R(OA_2 + OA_8)$$
$$OA_6 \cdot OA_7 = R(OA_1 - OA_4)$$

Suppose

$$OA_3 + OA_5 = M, \ \ OA_6 - OA_7 = N,$$
$$OA_2 + OA_8 = P, \ \ OA_1 - OA_4 = Q.$$

---

*The principal steps are given. For a full exposition see Catalan's *Théorèmes et Problèmes de Géométrie Élémentaire*. The treatment is given in full in Beman and Smith's translation cf Klein's *Famous Problems of Elementary Geometry*, chap. iv.

Then $MN = R^2$ and $PQ = R^2$

Again by substituting the values of $M$, $N$, $P$ and $Q$ in the formulas

$$MN = R^2, \quad PQ = R^2$$

and applying theorems i. and ii. we get

$$(M - N) - (P - Q) = R.$$

Also by substituting the values of $M$, $N$, $P$ and $Q$ in the above formula and applying theorems i. and ii. we get

$$(M - N)(P - Q) = 4R^2.$$

Hence $M - N$, $P - Q$, $M$, $N$, $P$ and $Q$ are determined.

Again

$$OA_2 + OA_8 = P,$$
$$OA_2 \cdot OA_8 = RN.$$

Hence $OA_8$ is determined.

**145.** By solving the equations we get

$$M - N = \tfrac{1}{2}R(1 + \sqrt{17}).$$

$$P - Q = \tfrac{1}{2}R(-1 + \sqrt{17}).$$

$$P = \tfrac{1}{4}R\left(-1 + \sqrt{17} + \sqrt{34 - 2\sqrt{17}}\right).$$

$$N = \tfrac{1}{4}R\left(-1 - \sqrt{17} + \sqrt{34 + 2\sqrt{17}}\right).$$

$$OA_8 = \tfrac{1}{8}R\Big[-1 + \sqrt{17} + \sqrt{34 - 2\sqrt{17}}$$
$$-2\sqrt{17 + 3\sqrt{17} + \sqrt{170 - 26\sqrt{17}} - 4\sqrt{34 + 2\sqrt{17}}}\,\Big]$$
$$= \tfrac{1}{8}R\Big[-1 + \sqrt{17} + \sqrt{34 - 2\sqrt{17}}$$
$$-2\sqrt{17 + 3\sqrt{17} - \sqrt{170 + 38\sqrt{17}}}\,\Big].$$

**146.** The geometric construction is as follows:

Let *BA* be the diameter of the given circle; *O* its center. Bisect *OA* in *C*. Draw *AD* at right angles to *OA* and take *AD = AB*. Draw *CD*. Take *E* and *E'* in *CD* and on each side of *C* so that *CE = CE' = CA*.

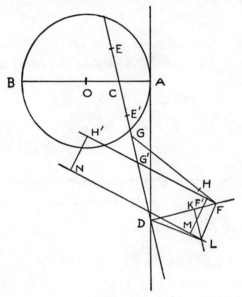

Fig. 51.

Bisect *ED* in *G* and *E'D* in *G'*. Draw *DF* perpendicular to *CD* and take *DF = OA*.

Draw *FG* and *FG'*.

Take *H* in *FG* and *H'* in *FG'* produced so that *GH = EG* and *G'H' = G'D*.

Then it is evident that

$$DE = M - N,$$
$$DE' = P - Q;$$

also

$$FH = N, \therefore (DE + FH) FH = DF^2 = R^2 ;$$
$$FH' = P, \therefore (FH' - DE') FH = DF^2 = R^2.$$

Again in $DF$ take $K$ such that $FK = FH$.

Draw $KL$ perpendicular to $DF$ and take $L$ in $KL$ such that $FL$ is perpendicular to $DL$.

Then $FL^2 = DF \cdot FK = RN$.

Again draw $H'N$ perpendicular to $FH'$ and take $H'N = FL$. Draw $NM$ perpendicular to $NH'$. Find $M$ in $NM$ such that $H'M$ is perpendicular to $FM$. Draw $MF'$ perpendicular to $FH'$.

Then

$$F'H' \cdot FF' = F'M^2 = FL^2$$
$$= RN.$$

But $FF' + F'H' = P.$

$$\therefore F'F = OA_8.$$

## XII. GENERAL PRINCIPLES.

**147.** In the preceding pages we have adopted several processes, e. g., bisecting and trisecting finite lines, bisecting rectilineal angles and dividing them into other equal parts, drawing perpendiculars to a given line, etc. Let us now examine the theory of these processes.

**148.** The general principle is that of congruence. Figures and straight lines are said to be congruent, if they are identically equal, or equal in all respects.

In doubling a piece of paper upon itself, we obtain the straight edges of two planes coinciding with each other. This line may also be regarded as the intersection of two planes if we consider their position during the process of folding.

In dividing a finite straight line, or an angle into a number of equal parts, we obtain a number of congruent parts. Equal lines or equal angles are congruent.

**149.** Let $X'X$ be a given finite line, divided into any two parts by $A'$. Take $O$ the mid-point by doubling the line on itself. Then $OA'$ is half the difference between $A'X$ and $X'A'$. Fold $X'X$ over $O$, and take $A$ in $OX$ corresponding to $A'$. Then $AA'$ is the difference between $A'X$ and $X'A'$ and it is bisected in $O$.

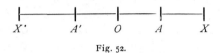

Fig. 52.

As $A'$ is taken nearer $O$, $A'O$ diminishes, and at the same time $A'A$ diminishes at twice the rate. This property is made use of in finding the mid-point of a line by means of the compasses.

**150.** The above observations apply also to an angle. The line of bisection is found easily by the compasses by taking the point of intersection of two circles.

**151.** In the line $X'X$, segments to the right of $O$ may be considered positive and segments to the eft of $O$ may be considered negative. That is, a point moving from $O$ to $A$ moves positively, and a point moving in the opposite direction $OA'$ moves negatively.

$$AX = OX - OA.$$
$$OA' = OX' - A'X',$$

both members of the equation being negative.*

**152.** If $OA$, one arm of an angle $AOP$, be fixed and $OP$ be considered to revolve round $O$, the angles which it makes with $OA$ are of different magnitudes.

*See Beman and Smith's *New Plane and Solid Geometry*, p. 56.

All such angles formed by *OP* revolving in the direction opposite to that of the hands of a watch are regarded positive. The angles formed by *OP* revolving in an opposite direction are regarded negative.*

**153.** After one revolution, *OP* coincides with *OA*. Then the angle described is called a perigon, which evidently equals four right angles. When *OP* has completed half the revolution, it is in a line with *OAB*. Then the angle described is called a straight angle, which evidently equals two right angles.†
When *OP* has completed quarter of a revolution, it is perpendicular to *OA*. All right angles are equal in magnitude. So are all straight angles and all perigons.

**154.** Two lines at right angles to each other form four congruent quadrants. Two lines otherwise inclined form four angles, of which those vertically opposite are congruent.

**155.** The position of a point in a plane is determined by its distance from each of two lines taken as above. The distance from one line is measured parallel to the other. In analytic geometry the properties of plane figures are investigated by this method. The two lines are called axes; the distances of the point from the axes are called co-ordinates, and the intersection of the axes is called the origin. This

---

* See Beman and Smith's *New Plane and Solid Geometry*, p. 56.
† *Ib.*, p. 5.

method was invented by Descartes in 1637 A. D.* It has greatly helped modern research.

**156.** If $X'X$, $YY'$ be two axes intersecting at $O$, distances measured in the direction of $OX$, i. e., to the right of $O$ are positive, while distances measured to the left of $O$ are negative. Similarly with reference to $YY'$, distances measured in the direction of $OY$ are positive, while distances measured in the direction of $OY'$ are negative.

**157.** Axial symmetry is defined thus: If two figures in the same plane can be made to coincide by turning the one about a fixed line in the plane through a straight angle, the two figures are said to be symmetric with regard to that line as axis of symmetry.†

**158.** Central symmetry is thus defined: If two figures in the same plane can be made to coincide by turning the one about a fixed point in that plane through a straight angle, the two figures are said to be symmetric with regard to that point as center of symmetry.‡

In the first case the revolution is outside the given plane, while in the second it is in the same plane.

If in the above two cases, the two figures are halves of one figure, the whole figure is said to be symmetric with regard to the axis or center—these are called axis or center of symmetry or simply axis or center.

* Beman and Smith's translation of Fink's *History of Mathematics*, p. 230.
† Beman and Smith's *New Plane and Solid Geometry*, p. 26.
‡ *Ib.*, p. 183.

**159.** Now, in the quadrant $XOY$ make a triangle $PQR$. Obtain its image in the quadrant $YOX'$ by folding on the axis $YY'$ and pricking through the paper at the vertices. Again obtain images of the two triangles in the fourth and third quadrants. It is seen that the triangles in adjacent quadrants posses axial

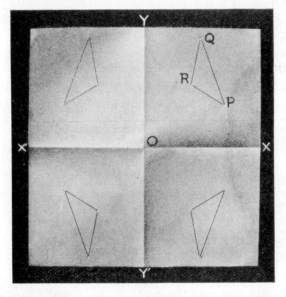

Fig. 53.

symmetry, while the triangles in alternate quadrants possess central symmetry.

**160.** Regular polygons of an odd number of sides possess axial symmetry, and regular polygons of an even number of sides possess central symmetry as well.

**161.** If a figure has two axes of symmetry at right angles to each other, the point of intersection of the axes is a center of symmetry. This obtains in regular polygons of an even number of sides and certain curves, such as the circle, ellipse, hyperbola, and the lemniscate; regular polygons of an odd number of

Fig. 54.

sides may have more axes than one, but no two of them will be at right angles to each other. If a sheet of paper is folded double and cut, we obtain a piece which has axial symmetry, and if it is cut fourfold, we obtain a piece which has central symmetry as well, as in Fig. 54.

**162.** Parallelograms have a center of symmetry. A quadrilateral of the form of a kite, or a trapezium with two opposite sides equal and equally inclined to either of the remaining sides, has an axis of symmetry.

**163.** The position of a point in a plane is also determined by its distance from a fixed point and the inclination of the line joining the two points to a fixed line drawn through the fixed point.

If $OA$ be the fixed line and $P$ the given point, the length $OP$ and $\angle AOP$, determine the position of $P$.

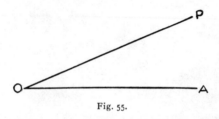

Fig. 55.

$O$ is called the pole, $OA$ the prime-vector, $OP$ the radius vector, and $\angle AOP$ the vectorial angle. $OP$ and $\angle AOP$ are called polar co-ordinates of $P$.

**164.** The image of a figure symmetric to the axis $OA$ may be obtained by folding through the axis $OA$. The radii vectores of corresponding points are equally inclined to the axis.

**165.** Let $ABC$ be a triangle. Produce the sides $CA$, $AB$, $BC$ to $D$, $E$, $F$ respectively. Suppose a person to stand at $A$ with face towards $D$ and then to proceed from *A* to *B*, *B* to *C*, and *C* to *A*. Then he successively describes the angles *DAB*, *EBC*, *FCD*. Having come to his original position *A*, he has com-

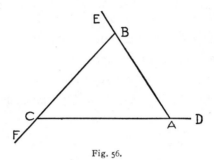

Fig. 56.

pleted a perigon, i. e., four right angles. We may therefore infer that the three exterior angles are together equal to four right angles.

The same inference applies to any convex polygon.

**161.** Suppose the man to stand at *A* with his face towards *C*, then to turn in the direction of *AB* and proceed along *AB*, *BC*, and *CA*.

In this case, the man completes a straight angle, i. e., two right angles. He successively turns through the angles *CAB*, *EBC*, and *FCA*. Therefore $\angle EBF + \angle FCA + \angle CAB$ (neg. angle) = a straight angle.

This property is made use of in turning engines on the railway. An engine standing upon *DA* with its head towards *A* is driven on to *CF*, with its head towards *F*. The motion is then reversed and it goes backwards to *EB*. Then it moves forward along *BA* on to *AD*. The engine has successively described the angles *ACB*, *CBA*, and *BAC*. Therefore the three interior angles of a triangle are together equal to two right angles.

**167.** The property that the three interior angles of a triangle are together equal to two right angles is illustrated as follows by paper folding.

Fold *CC″* perpendicular to *AB*. Bisect *C″B* in *N*, and *AC″* in *M*. Fold *NA′*, *MB′* perpendicular to *AB*, meeting *BC* and *AC* in *A′* and *B′*. Draw *A′C′*, *B′C′*.

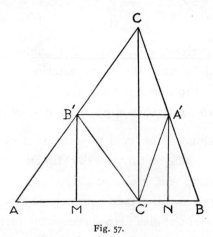

Fig. 57.

By folding the corners on *NA′*, *MB′* and *A′B′*, we find that the angles *A*, *B*, *C* of the triangle are equal to the angles *B′C′A*, *BC′A′*, and *A′C′B′* respectively, which together make up two right angles.

**168.** Take any line *ABC*. Draw perpendiculars to *ABC* at the points *A*, *B*, and *C*. Take points *D*, *E*, *F* in the respective perpendiculars equidistant from their feet. Then it is easily seen by superposition and proved by equal triangles that $DE$ is equal to $AB$ and perpendicular to $AD$ and $BE$, and that $EF$ is equal to $BC$ and perpendicular to $BE$ and $CF$. Now $AB$ $(=DE)$ is the shortest distance between the lines $AD$ and $BE$, and it is constant. Therefore $AD$

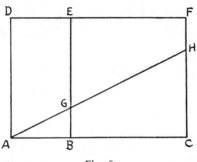

Fig. 58.

and $BE$ can never meet, i. e., they are parallel. Hence lines which are perpendicular to the same line are parallel.

The two angles $BAD$ and $EBA$ are together equal to two right angles. If we suppose the lines $AD$ and $BE$ to move inwards about $A$ and $B$, they will meet and the interior angles will be less than two right angles. They will not meet if produced backwards. This is embodied in the much abused twelfth postulate of Euclid's Elements.*

**169.** If $AGH$ be any line cutting $BE$ in $G$ and $CF$ in $H$, then

*For historical sketch see Beman and Smith's translation of Fink's *History of Mathematics*, p. 270.

∠ $GAD =$ the alternate ∠ $AGB$,

∴ each is the complement of ∠ $BAG$; and

∠ $HGE =$ the interior and opposite ∠ $GAD$.

∴ they are each $= ∠ AGB$.

Also the two angles $GAD$ and $EGA$ are together equal to two right angles.

**170.** Take a line $AX$ and mark off on it, from $A$, equal segments $AB$, $BC$, $CD$, $DE$....Erect perpendiculars to $AE$ at $B$, $C$, $D$, $E$....Let a line $AF'$ cut the perpendiculars in $B'$, $C'$, $D'$, $E'$....Then $AB'$, $B'C'$, $C'D'$, $D'E'$....are all equal.

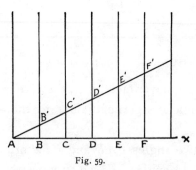

Fig. 59.

If $AB$, $BC$, $CD$, $DE$ be unequal, then

$$AB : BC = AB' : B'C'$$
$$BC : CD = B'C' : C'D',\ \text{and so on.}$$

**171.** If $ABCDE$....be a polygon, similar polygons may be obtained as follows.

Take any point $O$ within the polygon, and draw $OA$, $OB$, $OC$,....

Take any point $A'$ in $OA$ and draw $A'B'$, $B'C'$, $C'D'$,....parallel to $AB$, $BC$, $CD$....respectively.

Then the polygon $A'B'C'D'$....will be similar to $ABCD$.... The polygons so described around a common point are in perspective. The point $O$ may also lie outside the polygon. It is called the center of perspective.

**172.** To divide a given line into 2, 3, 4, 5....equal parts. Let $AB$ be the given line. Draw $AC$, $BD$

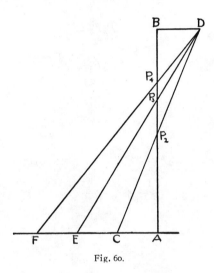

Fig. 60.

at right angles to $AB$ on opposite sides and make $AC = BD$. Draw $CD$ cutting $AB$ in $P_2$. Then $AP_2 = P_2B$.

Now produce $AC$ and take $CE = EF = FG$.... $= AC$ or $BD$. Draw $DE$, $DF$, $DG$....cutting $AB$ in $P_3$, $P_4$, $P_5$, ....

Then from similar triangles,

$$P_3B : AP_3 = BD : AE.$$

$$\therefore P_3B : AB = BD : AF$$
$$= 1 : 3.$$

Similarly

$$P_4B : AB = 1 : 4,$$

and so on.

If $AB = 1$,

$$AP_2 = \frac{1}{1\cdot2};$$

$$P_2P_3 = \frac{1}{2\cdot3};$$

$$P_3P_4 = \frac{1}{3\cdot4};$$

. . . . . . . . . . . . . . .

$$P_nP_{n+1} = \frac{1}{n(n+1)}.$$

But $AP_2 + P_2P_3 + P_3P_4 + \ldots$ is ultimately $= AB$.

$$\therefore \frac{1}{1\cdot2} + \frac{1}{2\cdot3} + \frac{1}{3\cdot4} + \ldots \text{ to } \infty = 1.$$

Or

$$1 - \frac{1}{2} = \frac{1}{1\cdot2};$$

$$\frac{1}{2} - \frac{1}{3} = \frac{1}{2\cdot3};$$

. . . . . . . . . . . . . . . .

$$\frac{1}{n} - \frac{1}{n+1} = \frac{1}{n(n+1)}.$$

Adding

$$\frac{1}{1\cdot2} + \frac{1}{2\cdot3} + \ldots + \frac{1}{n(n+1)} = 1 - \frac{1}{n+1}.$$

$$\therefore \ \frac{1}{1\cdot 2} + \frac{1}{2\cdot 3} \cdots + \frac{1}{(n-1)n} = 1 - \frac{1}{n}.$$

The limit of $1 - \dfrac{1}{n}$ when $n$ is $\infty$ is 1.

**173.** The following simple contrivance may be used for dividing a line into a number of equal parts.

Take a rectangular piece of paper, and mark off $n$ equal segments on each or one of two adjacent sides. Fold through the points of section so as to obtain perpendiculars to the sides. Mark the points of section and the corners 0, 1, 2, . . . . $n$. Suppose it is required to divide the edge of another piece of paper $AB$ into $n$ equal parts. Now place $AB$ so that $A$ or $B$ may lie on 0, and $B$ or $A$ on the perpendicular through $n$.

In this case $AB$ must be greater than $ON$. But the smaller side of the rectangle may be used for smaller lines.

The points where $AB$ crosses the perpendiculars are the required points of section.

**174.** Center of mean position. If a line $AB$ contains $(m+n)$ equal parts, and it is divided at $C$ so that $AC$ contains $m$ of these parts and $CB$ contains $n$ of them ; then if from the points $A$, $C$, $B$ perpendiculars $AD$, $CF$, $BE$ be let fall on any line,

$$m\cdot BE + n\cdot AD = (m+n)\cdot CF.$$

Now, draw $BGH$ parallel to $ED$ cutting $CF$ in $G$ and $AD$ in $H$. Suppose through the points of division $AB$ lines are drawn parallel to $BH$. These lines will divide $AH$ into $(m + n)$ equal parts and $CG$ into $n$ equal parts.

$$\therefore \; n \cdot AH = (m + n) \cdot CG,$$

and since $DH$ and $BE$ are each $= GF$,

$$n \cdot HD + m \cdot BE = {}^{\prime}m + n)GF.$$

Hence, by addition

$$n \cdot 1 + D + m \cdot BE = (m + n)\,GF$$

$$n \cdot AD + m \cdot BE = (m + n) \cdot CF.$$

$C$ is called the center of mean position, or the mean center of $A$ and $B$ for the system of multiples $m$ and $n$.

The principle can be extended to any number of points, not in a line. Then if $P$ represent the feet of the perpendiculars on any line from $A$, $B$, $C$, etc., if $a$, $b$, $c$ ...be the corresponding multiples, and if $M$ be the mean center

$$a \cdot AP + b \cdot BP + c \cdot CP \ldots$$

$$= (a + b + c + \ldots) \cdot MP.$$

If the multiples are all equal to $a$, we get

$$a(AP + BP + CP + \ldots) = na \cdot MP$$

$n$ being the number of points.

**175.** The center of mean position of a number of points with equal multiples is obtained thus. Bisect the line joining any two points $A$, $B$ in $G$, join $G$ to a third point $C$ and divide $GC$ in $H$ so that $GH = \frac{1}{3}GC$; join $H$ to a fourth point $D$ and divide $HD$ in $K$ so that $HK = \frac{1}{4}HD$ and so on: the last point found will be the center of mean position of the system of points.

**176.** The notion of mean center or center of mean position is derived from Statics, because a system of material points having their weights denoted by $a,$ $b,$ $c\ldots$, and placed at $A,$ $B,$ $C\ldots$ would balance about the mean center $M,$ if free to rotate about $M$ under the action of gravity.

The mean center has therefore a close relation to the center of gravity of Statics.

**177.** The mean center of three points not in a line, is the point of intersection of the medians of the triangle formed by joining the three points. This is also the center of gravity or mass center of a thin triangular plate of uniform density.

**178.** If $M$ is the mean center of the points $A,$ $B,$ $C,$ etc., for the corresponding multiples $a,$ $b,$ $c,$ etc., and if $P$ is any other point, then

$$a \cdot AP^2 + b \cdot BP^2 + c \cdot CP^2 + \ldots$$
$$= a \cdot AM^2 + b \cdot BM^2 + c \cdot CM^2 + \ldots$$
$$+ PM^2 (a + b + c + \ldots).$$

Hence in any regular polygon, if $O$ is the in-center or circum-center and $P$ is any point

$$AP^2 + BP^2 + \ldots = OA^2 + OB^2 + \ldots + n \cdot OP^2$$
$$= n \cdot (R^2 + OP^2).$$

Now

$$AB^2 + AC^2 + AD^2 + \ldots = 2n \cdot R^2.$$

Similarly

$$BA^2 + BC^2 + BD^2 + \ldots = 2n \cdot R^2$$
$$CA^2 + CB^2 + CD^2 + \ldots = 2n \cdot R^2.$$

Adding

$$2(AB^2 + AC^2 + AD^2 + \ldots) = n \cdot 2n \cdot R^2.$$
$$\therefore \quad AB^2 + AC^2 + AD^2 + \ldots = n^2 \cdot R^2.$$

**179.** The sum of the squares of the lines joining the mean center with the points of the system is a minimum.

If $M$ be the mean center and $P$ any other point not belonging to the system,

$\Sigma PA^2 = \Sigma MA^2 + \Sigma PM^2$, (where $\Sigma$ stands for "the sum of all expressions of the type").

$\therefore \ \Sigma PA^2$ is the minimum when $PM = 0$, i. e., when $P$ is the mean center.

**180.** Properties relating to concurrence of lines and collinearity of points can be tested by paper folding.* Some instances are given below :

(1) The medians of a triangle are concurrent. The common point is called the centroid.

(2) The altitudes of a triangle are concurrent. The common point is called the orthocenter.

(3) The perpendicular bisectors of the sides of a triangle are concurrent. The common point is called the circum-center.

(4) The bisectors of the angles of a triangle are concurrent. The common point is called the in-center.

(5) Let $ABCD$ be a parallelogram and $P$ any point. Through $P$ draw $GH$ and $EF$ parallel to $BC$

---

*For treatment of certain of these properties see Beman and Smith's *New Plane and Solid Geometry*, pp. 84, 182.

and *AB* respectively. Then the diagonals *EG*, *HF*, and the line *DB* are concurrent.

(6) If two similar unequal rectineal figures are so placed that their corresponding sides are parallel, then the joins of corresponding corners are concurrent. The common point is called the center of similarity.

(7) If two triangles are so placed that their corners are two and two on concurrent lines, then their corresponding sides intersect collinearly. This is known as Desargues's theorem. The two triangles are said to be in perspective. The point of concurrence and line of collinearity are respectively called the center and axis of perspective.

(8) The middle points of the diagonals of a complete quadrilateral are collinear.

(9) If from any point on the circumference of the circum-circle of a triangle, perpendiculars are dropped on its sides, produced when necessary, the feet of these perpendiculars are collinear. This line is called Simson's line.

Simson's line bisects the join of the orthocenter and the point from which the perpendiculars are drawn.

(10) In any triangle the orthocenter, circum-center, and centroid are collinear.

The mid-point of the join of the orthocenter and circum-center is the center of the nine-points circle, so called because it passes through the feet of the altitudes and medians of the triangle and the mid-point of that part of each altitude which lies between the orthocenter and vertex.

The center of the nine-points circle is twice as far from the orthocenter as from the centroid. This is known as Poncelet's theorem.

(11) If *A, B, C, D, E, F,* are any six points on a circle which are joined successively in any order, then the intersections of the first and fourth, of the second and fifth, and of the third and sixth of these joins produced when necessary) are collinear. This is known as Pascal's theorem.

(12) The joins of the vertices of a triangle with the points of contact of the in-circle are concurrent. The same property holds for the ex circles.

(13) The internal bisectors of two angles of a triangle, and the external bisector of the third angle intersect the opposite sides collinearly.

(14) The external bisectors of the angles of a triangle intersect the opposite sides collinearly.

(15) If any point be joined to the vertices of a triangle, the lines drawn through the point perpendicular to those joins intersect the opposite sides of the triangle collinearly.

(16) If on an axis of symmetry of the congruent triangles *ABC, A'B'C'* a point *O* be taken *A'O, B'O,* and *C'O* intersect the sides *BC, CA* and *AB* collinearly.

(17) The points of intersection of pairs of tangents to a circle at the extremities of chords which pass

through a given point are collinear. This line is called the polar of the given point with respect to the circle.

(18) The isogonal conjugates of three concurrent lines $AX$, $BX$, $CX$ with respect to the three angles of a triangle $ABC$ are concurrent. (Two lines $AX$, $AY$ are said to be isogonal conjugates with respect to an angle $BAC$, when they make equal angles with its bisector.)

(19) If in a triangle $ABC$, the lines $AA'$, $BB'$, $CC'$ drawn from each of the angles to the opposite sides are concurrent, their isotomic conjugates with respect to the corresponding sides are also concurrent. (The lines $AA'$, $AA''$ are said to be isotomic conjugates, with respect to the side $BC$ of the triangle $ABC$, when the intercepts $BA'$ and $CA''$ are equal.)

(20) The three symmedians of a triangle are concurrent. (The isogonal conjugate of a median $AM$ of a triangle is called a symmedian.)

# XIII. THE CONIC SECTIONS.

## SECTION I.—THE CIRCLE.

**181.** A piece of paper can be folded in numerous ways through a common point. Points on each of the lines so taken as to be equidistant from the common point will lie on the circumference of a circle, of which the common point is the center. The circle is the locus of points equidistant from a fixed point, the centre.

**182.** Any number of concentric circles can be drawn. They cannot meet each other.

**183.** The center may be considered as the limit of concentric circles described round it as center, the radius being indefinitely diminished.

**184.** Circles with equal radii are congruent and equal.

**185.** The curvature of a circle is uniform throughout the circumference. A circle can therefore be made to slide along itself by being turned about its center. Any figure connected with the circle may be turned about the center of the circle without changing its relation to the circle.

**186.** A straight line can cross a circle in only two points.

**187.** Every diameter is bisected at the center of the circle. It is equal in length to two radii. All diameters, like the radii, are equal.

**188.** The center of a circle is its center of symmetry, the extremities of any diameter being corresponding points.

**189.** Every diameter is an axis of symmetry of the circle, and conversely.

**190.** The propositions of §§ 188, 189 are true for systems of concentric circles.

**191.** Every diameter divides the circle into two equal halves called semicircles.

**192.** Two diameters at right angles to each other divide the circle into four equal parts called quadrants.

**193.** By bisecting the right angles contained by the diameters, then the half right angles, and so on, we obtain $2^n$ equal sectors of the circle. The angle between the radii of each sector is $\dfrac{4}{2^n}$ of a right angle or $\dfrac{2\pi}{2^n} = \dfrac{\pi}{2^{n-1}}$.

**194.** As shown in the preceding chapters, the right angle can be divided also into 3, 5, 9, 10, 12, 15 and 17 equal parts. And each of the parts thus obtained can be subdivided into $2^n$ equal parts.

**195.** A circle can be inscribed in a regular polygon, and a circle can also be circumscribed round it. The former circle will touch the sides at their mid-points.

**196.** Equal arcs subtend equal angles at the center; and conversely. This can be proved by superposition. If a circle be folded upon a diameter, the two semicircles coincide. Every point in one semicircumference has a corresponding point in the other, below it.

**197.** Any two radii are the sides of an isosceles triangle, and the chord which joins their extremities is the base of the triangle.

**198.** A radius which bisects the angle between two radii is perpendicular to the base chord and also bisects it.

**199.** Given one fixed diameter, any number of pairs of radii may be drawn, the two radii of each set being equally inclined to the diameter on each side of it. The chords joining the extremities of each pair of radii are at right angles to the diameter. The chords are all parallel to one another.

**200.** The same diameter bisects all the chords as well as arcs standing upon the chords, i. e., the locus of the mid-points of a system of parallel chords is a diameter.

**201.** The perpendicular bisectors of all chords of a circle pass through the center.

**202.** Equal chords are equidistant from the center.

**203.** The extremities of two radii which are equally inclined to a diameter on each side of it, are equidistant from every point in the diameter. Hence, any number of circles can be described passing through the two points. In other words, the locus of the centers of circles passing through two given points is the straight line which bisects at right angles the join of the points.

**204.** Let $CC'$ be a chord perpendicular to the radius $OA$. Then the angles $AOC$ and $AOC'$ are equal. Suppose both move on the circumference towards $A$ with the same velocity, then the chord $CC'$ is always parallel to itself and perpendicular to $OA$. Ultimately the points $C$, $A$ and $C'$ coincide at $A$, and $CAC'$ is perpendicular to $OA$. $A$ is the last point common to the chord and the circumference. $CAC'$ produced becomes ultimately a tangent to the circle.

**205.** The tangent is perpendicular to the diameter through the point of contact; and conversely.

**206.** If two chords of a circle are parallel, the arcs joining their extremities towards the same parts are equal. So are the arcs joining the extremities of either chord with the diagonally opposite extremities of the other and passing through the remaining extremities. This is easily seen by folding on the diameter perpendicular to the parallel chords.

**207.** The two chords and the joins of their extremities towards the same parts form a trapezoid which has an axis of symmetry, viz., the diameter perpendicular to the parallel chords. The diagonals of the trapezoid intersect on the diameter. It is evident by folding that the angles between each of the parallel chords and each diagonal of the trapezoid are equal. Also the angles upon the other equal arcs are equal.

**208.** The angle subtended at the center of a circle by any arc is double the angle subtended by it at the circumference.

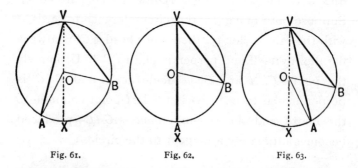

Fig. 61.          Fig. 62.          Fig. 63.

An inscribed angle equals half the central angle standing on the same arc.

Given

$AVB$ an inscribed angle, and $AOB$ the central angle on the same arc $AB$.

To prove    that $\angle AVB = \frac{1}{2} \angle AOB$.

Proof.

1. Suppose $VO$ drawn through center $O$, and produced to meet the circumference at $X$.

Then $\angle XVB = \angle VBO$.

2. And $\quad \angle XOB = \angle XVB + \angle VBO$,

$\qquad = 2 \angle XVB$.

3. $\quad \therefore \angle XVB = \frac{1}{2} \angle XOB$.

4. Similarly $\angle AVX = \frac{1}{2} \angle AOX$ (each=zero in Fig. 62),

and $\therefore \angle AVB = \frac{1}{2} \angle AOB$.

The proof holds for all three figures, point $A$ having moved to $X$ (Fig. 62), and then through $X$ (Fig. 63).*

**209.** The angle at the center being constant, the angles subtended by an arc at all points of the circumference are equal.

**210.** The angle in a semicircle is a right angle.

**211.** If $AB$ be a diameter of a circle, and $DC$ a chord at right angles to it, then $ACBD$ is a quadrilateral of which $AB$ is an axis of symmetry. The angles $BCA$ and $ADB$ being each a right angle, the remaining two angles $DBC$ and $CAD$ are together equal to a straight angle. If $A'$ and $B'$ be any other points on the arcs $DAC$ and $CBD$ respectively, the $\angle CAD = \angle CA'D$ and $\angle DBC = \angle DB'C$, and $\angle CA'D + DB'C =$ a straight angle. Therefore, also, $\angle B'CA' + \angle A'DB' =$ a straight angle.

Conversely, if a quadrilateral has two of its opposite angles together equal to two right angles, it is inscriptible in a circle.

---

*The above figures and proof are from Beman and Smith's *New Plane and Solid Geometry*, p. 129.

**212.** The angle between the tangent to a circle and a chord which passes through the point of contact is equal to the angle at the circumference standing upon that chord and having its vertex on the side of it opposite to that on which the first angle lies.

Let $AC$ be a tangent to the circle at $A$ and $AB$ a chord. Take $O$ the center of the circle and draw $OA$, $OB$. Draw $OD$ perpendicular to $AB$.

Then $\angle BAC = \angle AOD = \frac{1}{2} \angle BOA$.

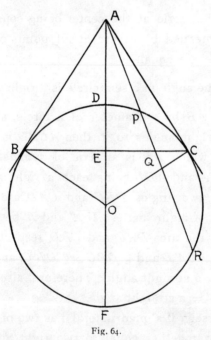

Fig. 64.

**213.** Perpendiculars to diameters at their extremities touch the circle at these extremities. (See Fig. 64). The line joining the center and the point of intersection of two tangents bisects the angles between the two tangents and between the two radii. It also bisects the join of the points of contact. The tangents are equal.

This is seen by folding through the center and the point of intersection of the tangents.

Let $AC$, $AB$ be two tangents and $ADEOF$ the line through the intersection of the tangents $A$ and the center $O$, cutting the circle in $D$ and $F$ and $BC$ in $E$.

Then $AC$ or $AB$ is the geometric mean of $AD$ and $AF$; $AE$ is the harmonic mean; and $AO$ the arithmetic mean.

$$AB^2 = AD \cdot AF.$$
$$AB^2 = OA \cdot AE.$$
$$\therefore AE = \frac{AD \cdot AF}{OA} = \frac{2AD \cdot AF}{AD + AF}.$$

Similarly, if any other chord through $A$ be obtained cutting the circle in $P$ and $R$ and $BC$ in $Q$, then $AQ$ is the harmonic mean and $AC$ the geometric mean between $AP$ and $AR$.

214. Fold a right-angled triangle $OCB$ and $CA$ the perpendicular on the hypotenuse. Take $D$ in $AB$ such that $OD = OC$ (Fig. 65).

$$\text{Then} \quad OA \cdot OB = OC^2 = OD^2,$$
$$\text{and} \quad OA : OC = OC : OB,$$
$$OA : OD = OD : OB.$$

A circle can be described with $O$ as center and $OC$ or $OD$ as radius.

The points $A$ and $B$ are inverses of each other with reference to the center of inversion $O$ and the circle of inversion $CDE$.

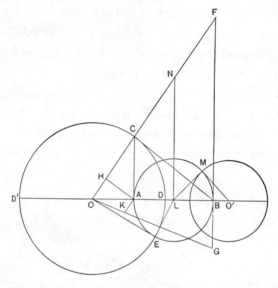

Fig. 65.

Hence when the center is taken as the origin, the foot of the ordinate of a point on a circle has for its inverse the point of intersection of the tangent and the axis taken.

**215.** Fold $FBG$ perpendicular to $OB$. Then the line $FBG$ is called the polar of point $A$ with reference to the polar circle $CDE$ and polar center $O$; and $A$ is called the pole of $FBG$. Conversely $B$ is the pole of

*CA* and *CA* is the polar of *B* with reference to the same circle.

**216.** Produce *OC* to meet *FBG* in *F*, and fold *AH* perpendicular to *OC*.

Then *F* and *H* are inverse points.

*AH* is the polar of *F*, and the perpendicular at *F* to *OF* is the polar of *H*.

**217.** The points *A*, *B*. *F*, *H*, are concyclic.

That is, two points and their inverses are concyclic ; and conversely.

Now take another point *G* on *FBG*. Draw *OG*, and fold *AK* perpendicular to *OG*. Then *K* and *G* are inverse points with reference to the circle *CDE*.

**218.** The points *F*, *B*, *G* are collinear, while their polars pass through *A*.

Hence, the polars of collinear points are concurrent.

**219.** Points so situated that each lies on the polar of the other are called conjugate points, and lines so related that each passes through the pole of the other are called conjugate lines.

*A* and *F* are conjugate points, so are *A* and *B*, *A* and *G*.

The point of intersection of the polars of two points is the pole of the join of the points.

**220.** As *A* moves towards *D*, *B* also moves up to it. Finally *A* and *B* coincide and *FBG* is the tangent at *B*.

Hence the polar of any point on the circle is the tangent at that point.

**221.** As $A$ moves back to $O$, $B$ moves forward to infinity. The polar of the center of inversion or the polar center is the line at infinity.

**222.** The angle between the polars of two points is equal to the angle subtended by these points at the polar center.

**223.** The circle described with $B$ as a center and $BC$ as a radius cuts the circle $CDE$ orthogonally.

**224.** Bisect $AB$ in $L$ and fold $LN$ perpendicular to $AB$. Then all circles passing through $A$ and $B$ will have their centers on this line. These circles cut the circle $CDE$ orthogonally. The circles circumscribing the quadrilaterals $ABFH$ and $ABGK$ are such circles. $AF$ and $AG$ are diameters of the respective circles. Hence if two circles cut orthogonally the extremities of any diameter of either are conjugate points with respect to the other.

**225.** The points $O$, $A$, $H$ and $K$ are concyclic. $H$, $A$, $K$ being inverses of points on the line $FBG$, the inverse of a line is a circle through the center of inversion and the pole of the given line, these points being the extremities of a diameter; and conversely.

**226.** If $DO$ produced cuts the circle $CDE$ in $D'$, $D$ and $D'$ are harmonic conjugates of $A$ and $B$. Similarly, if any line through $B$ cuts $AC$ in $A'$ and the circle $CDE$ in $d$ and $d''$, then $d$ and $d''$ are harmonic conjugates of $A'$ and $B$.

**227.** Fold any line $LM = LB = LA$, and $MO'$ perpendicular to $LM$ meeting $AB$ produced in $O'$.

Then the circle described with center $O'$ and radius $O'M$ cuts orthogonally the circle described with center $L$ and radius $LM$.

Now $$OL^2 = OE^2 + LE^2,$$
and $$O'L^2 = O'M^2 + LM^2.$$

$$\therefore \; OL^2 - O'L^2 = OE^2 - O'M^2.$$

$\therefore$ $LN$ is the radical axis of the circles $O\,(OC)$ and $O'(O'M)$.

By taking other points in the semicircle $AMB$ and repeating the same construction as above, we get two infinite systems of circles co-axial with $O(OC)$ and $O'(O'M)$, viz., one system on each side of the radical axis, $LN$. The point circle of each system is a point, $A$ or $B$, which may be regarded as an infinitely small circle.

The two infinite systems of circles are to be regarded as one co-axial system, the circles of which range from infinitely large to infinitely small—the radical axis being the infinitely large circle, and the limiting points the infinitely small. This system of co-axial circles is called the limiting point species.

If two circles cut each other their common chord is their radical axis. Therefore all circles passing through $A$ and $B$ are co-axial. This system of co-axial circles is called the common point species.

**228.** Take two lines $OAB$ and $OPQ$. From two points $A$ and $B$ in $OAB$ draw $AP$, $BQ$ perpendicular to $OPQ$. Then circles described with $A$ and $B$ as centers and $AP$ and $BQ$ as radii will touch the line $OPQ$ at $P$ and $Q$.

Then                $OA : OB = AP : BQ$.

This holds whether the perpendiculars are towards the same or opposite parts. The tangent is in one case direct, and in the other transverse.

In the first case, $O$ is outside $AB$, and in the second it is between $A$ and $B$. In the former it is called the external center of similitude and in the latter the internal centre of similitude of the two circles.

**229.** The line joining the extremities of two parallel radii of the two circles passes through their external center of similitude, if the radii are in the same direction, and through their internal center, if they are drawn in opposite directions.

**230.** The two radii of one circle drawn to its points of intersection with any line passing through either center of similitude, are respectively parallel to the two radii of the other circle drawn to its intersections with the same line.

**231.** All secants passing through a center of similitude of two circles are cut in the same ratio by the circles.

**232..** If $B_1$, $D_1$, and $B_2$, $D_2$ be the points of inter-section, $B_1$, $B_2$, and $D_1$, $D_2$ being corresponding points,

$$OB_1 \cdot OD_2 = OD_1 \cdot OB_2 = OC_2{}^2 \cdot \frac{X_1 C_1}{X_2 C_2}.$$

Hence the inverse of a circle, not through the center of inversion is a circle.

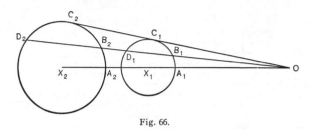

Fig. 66.

The center of inversion is the center of similitude of the original circle and its inverse.

The original circle, its inverse, and the circle of inversion are co-axial.

**233.** The method of inversion is one of the most important in the range of Geometry. It was discovered jointly by Doctors Stubbs and Ingram, Fellows of Trinity College, Dublin, about 1842. It was employed by Sir William Thomson in giving geometric proof of some of the most difficult propositions in the mathematical theory of electricity.

### SECTION II.—THE PARABOLA.

**234.** A parabola is the curve traced by a point which moves in a plane in such a manner that its distance from a given point is always equal to its distance from a given straight line.

**235.** Fig. 67 shows how a parabola can be marked on paper. The edge of the square *MN* is the directrix, *O* the vertex, and *F* the focus. Fold through *OX* and obtain the axis. Divide the upper half of the

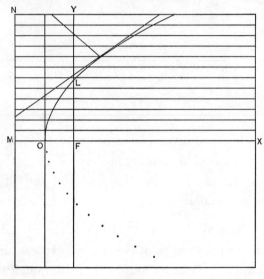

Fig. 67.

square into a number of sections by lines parallel to the axis. These lines meet the directrix in a number of points. Fold by laying each of these points on the focus and mark the point where the corresponding horizontal line is cut. The points thus obtained lie on a parabola. The folding gives also the tangent to the curve at the point.

**236.** *FL* which is at right angles to *OX* is called the semi-latus rectum.

**237.** When points on the upper half of the curve have been obtained, corresponding points on the lower half are obtained by doubling the paper on the axis and pricking through them.

**238.** When the axis and the tangent at the vertex are taken as the axes of co-ordinates, and the vertex as origin, the equation of the parabola becomes

$$y^2 = 4ax \text{ or } PN^2 = 4 \cdot OF \cdot ON.$$

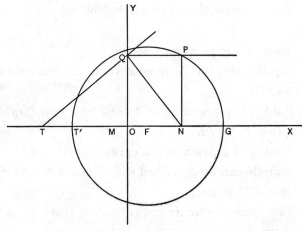

Fig. 68.

The parabola may be defined as the curve traced by a point which moves in one plane in such a manner that the square of its distance from a given straight line varies as its distance from another straight line; or the ordinate is the mean proportional between the abscissa, and the latus rectum which is equal to $4 \cdot OF$. Hence the following construction.

Take $OT$ in $FO$ produced $= 4 \cdot OF$.

Bisect $TN$ in $M$.

Take $Q$ in $OY$ such that $MQ = MN = MT$.

Fold through $Q$ so that $QP$ may be at right angles to $OY$.

Let $P$ be the point where $QP$ meets the ordinate of $N$.

Then $P$ is a point on the curve.

**239.** The subnormal $= 2OF$ and $FP = FG = FT''$.

These properties suggest the following construction.

Take $N$ any point on the axis.

On the side of $N$ remote from the vertex take $NG = 2OF$.

Fold $NP$ perpendicular to $OG$ and find $P$ in $NP$ such that $FP = FG$.

Then $P$ is a point on the curve.

A circle can be described with $F$ as center and $FG$, $FP$ and $FT'$ as radii.

The double ordinate of the circle is also the double ordinate of the parabola, i. e., $P$ describes a parabola as $N$ moves along the axis.

**240.** Take any point $N'$ between $O$ and $F$ (Fig. 69). Fold $RN'P'$ at right angles to $OF$.

Take $R$ so that $OR = OF$.

Fold $RN$ perpendicular to $OR$, $N$ being on the axis.

Fold *NP* perpendicular to the axis.

Now, in *OX* take *OT = ON'*.

Take *P'* in *RN'* so that *FP' = FT*.

Fold through *P'F* cutting *NP* in *P*.

Then *P* and *P'* are points on the curve.

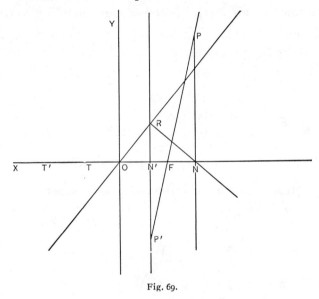

Fig. 69.

**241.** *N* and *N'* coincide when *PFP'* is the latus rectum.

As *N'* recedes from *F* to *O*, *N* moves forward from *F* to infinity.

At the same time, *T* moves toward *O*, and *T'*(*OT'* = *ON*) moves in the opposite direction toward infinity.

**242.** To find the area of a parabola bounded by the axis and an ordinate.

Complete the rectangle *ONPK*. Let *OK* be divided into $n$ equal portions of which suppose $Om$ to contain $r$ and $mn$ to be the $(r+1)^{th}$. Draw $mp$, $nq$ at right angles to $OK$ meeting the curve in $p$, $q$, and $pn'$ at right angles to $nq$. The curvilinear area $OPK$ is the limit of the sum of the series of rectangles constructed as $mn'$ on the portions corresponding to $mn$.

But $\square pn : \square NK = pm \cdot mn : PK \cdot OK$,

and, by the properties of the parabola,

$$pm : PK = Om^2 : OK^2$$
$$= r^2 : n^2$$

and $mn : OK = 1 : n$.

$$\therefore pm \cdot mn : PK \cdot OK = r^2 : n^3.$$

$$\therefore \square pn = \frac{r^2}{n^3} \times \square NK.$$

Hence the sum of the series of rectangles

$$= \frac{1^2 + 2^2 + 3^2 \ldots + (n-1)^2}{n^3} \times \square NK$$

$$= \frac{(n-1)\,n\,(2n-1)}{1 \cdot 2 \cdot 3 \cdot n^3} \times \square NK$$

$$= \frac{2n^3 - 3n^2 + n}{1 \cdot 2 \cdot 3 \cdot n^3} \times \square NK$$

$$= \left( \frac{1}{3} - \frac{1}{2n} + \frac{1}{6n^2} \right) \times \square NK$$

$$= \tfrac{1}{3} \text{ of } \square NK \text{ in the limit, i. e., when } n \text{ is } \infty.$$

$\therefore$ The curvilinear area $OPK = \tfrac{1}{3}$ of $\square NK$, and the parabolic area $OPN = \tfrac{2}{3}$ of $\square NK$.

**243.** The same line of proof applies when any diameter and an ordinate are taken as the boundaries of the parabolic area.

### SECTION III.—THE ELLIPSE.

**244.** An ellipse is the curve traced by a point which moves in a plane in such a manner that its distance from a given point is in a constant ratio of less inequality to its distance from a given straight line.

Let *F* be the focus, *OY* the directrix, and *XX'* the perpendicular to *OY* through *F*. Let *FA : AO* be the

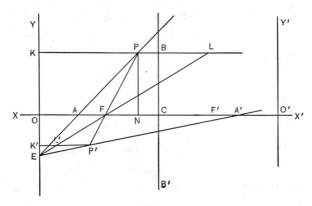

Fig. 70.

constant ratio, *FA* being less than *AO*. *A* is a point on the curve called the vertex.

As in § 116, find *A'* in *XX'* such that

$$FA' : A'O = FA : AO.$$

Then *A'* is another point on the curve, being a second vertex.

Double the line *AA'* on itself and obtain its middle point *C*, called the center, and mark *F'* and *O'* corresponding to *F* and *O*. Fold through *O'* so that *O'Y'* may be at right angles to $XX'$.   Then $F'$ is the second focus and $O'Y'$ the second directrix.

By folding $AA'$, obtain the perpendicular through $C$.

$$FA : AO = FA' : A'O$$
$$= FA + FA' : AO + A'O$$
$$= AA' : OO'$$
$$= CA : CO.$$

Take points $B$ and $B'$ in the perpendicular through $C$ and on opposite sides of it, such that $FB$ and $FB'$ are each equal to $CA$.   Then $B$ and $B'$ are points on the curve.

$AA'$ is called the major axis, and $BB'$ the minor axis.

**245.** To find other points on the curve, take any point $E$ in the directrix, and fold through $E$ and $A$, and through $E$ and $A'$.   Fold again through $E$ and $F$ and mark the point $P$ where $FA'$ cuts $EA$ produced. Fold through $PF$ and $P'$ on $EA'$.   Then $P$ and $P'$ are points on the curve.

Fold through $P$ and $P'$ so that $KPL$ and $K'L'F'$ are perpendicular to the directrix, $K$ and $K'$ being on the directrix and $L$ and $L'$ on $EL$.

$FL$ bisects the angle $A'FP$,

$$\therefore \quad \angle LFP = \angle PLF \text{ and } FP = PL.$$
$$FP : PK = PL : PK$$
$$= FA : AO.$$

And

$$FP' : P'K' = P'L' : P'K'$$

$$= FA' : A'O$$
$$= FA : AO.$$

If $EO = FO$, $FP$ is at right angles to $FO$, and $FP = FP'$. $PP'$ is the latus rectum.

**246.** When a number of points on the left half of the curve are found, corresponding points on the other half can be marked by doubling the paper on the minor axis and pricking through them.

**247.** An ellipse may also be defined as follows :

If a point $P$ move in such a manner that $PN^2 : AN \cdot NA'$ is a constant ratio, $PN$ being the distance of $P$ from the line joining two fixed points $A$, $A'$, and $N$ being between $A$ and $A'$, the locus of $P$ is an ellipse of which $AA'$ is an axis.

**248.** In the circle, $PN^2 = AN \cdot NA'$.

In the ellipse $PN^2 : AN \cdot NA'$ is a constant ratio.

This ratio may be less or greater than unity. In the former case $\angle APA'$ is obtuse, and the curve lies within the auxiliary circle described on $AA'$ as diameter. In the latter case, $\angle APA'$ is acute and the curve is outside the circle. In the first case $AA'$ is the major, and in the second it is the minor axis.

**249.** The above definition corresponds to the equation

$$y^2 = \frac{b^2}{a^2} (2ax - x^2)$$

when the vertex is the origin.

**250.** $AN \cdot NA'$ is equal to the square on the ordinate $QN$ of the auxiliary circle, and $PN : QN = BC : AC$.

**251.** Fig. 71 shows how the points can be determined when the constant ratio is less than unity. Thus, lay off $CD = AC$, the semi-major axis. Through $E$ any point of $AC$ draw $DE$ and produce it to meet

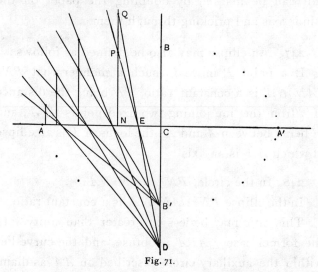

Fig. 71.

the auxiliary circle in $Q$. Draw $B'E$ and produce it to meet the ordinate $QN$ in $P$. Then is $PN : QN = B'C : DC = BC : AC$. The same process is applicable when the ratio is greater than unity. When points in one quadrant are found, corresponding points in other quadrants can be easily marked.

**252.** If $P$ and $P'$ are the extremities of two conjugate diameters of an ellipse and the ordinates $MP$

and $M'P'$ meet the auxiliary circle in $Q$ and $Q'$, the angle $QCQ'$ is a right angle.

Now take a rectangular piece of card or paper and mark on two adjacent edges beginning with the common corner lengths equal to the minor and major axes. By turning the card round $C$ mark corresponding points on the outer and inner auxiliary circles. Let $Q$, $R$ and $Q'$, $R'$ be the points in one position. Fold the ordinates $QM$ and $Q'M'$, and $RP$ and $R'P'$, perpendiculars to the ordinates. Then $P$ and $P'$ are points on the curve.

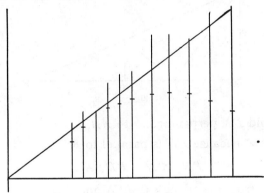

Fig. 72.

**253.** Points on the curve may also be easily determined by the application of the following property of the conic sections.

The focal distance of a point on a conic is equal

to the length of the ordinate produced to meet the tangent at the end of the latus rectum.

**254.** Let $A$ and $A'$ be any two points. Draw $AA'$ and produce the line both ways. From any point $D$ in $A'A$ produced draw $DR$ perpendicular to $AD$. Take any point $R$ in $DR$ and draw $RA$ and $RA'$. Fold $AP$ perpendicular to $AR$, meeting $RA'$ in $P$. For different positions of $R$ in $DR$, the locus of $P$ is an ellipse, of which $AA'$ is the major axis.

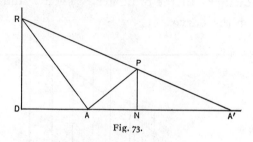

Fig. 73.

Fold $PN$ perpendicular to $AA'$.

Now, because $PN$ is parallel to $RD$,

$$PN : A'N = RD : A'D.$$

Again, from the triangles, $APN$ and $DAR$,

$$PN : AN = AD : RD.$$

$\therefore PN^2 : AN \cdot A'N = AD : A'D$, a constant ratio, less than unity, and it is evident from the construction that $N$ must lie between $A$ and $A'$.

### SECTION IV.—THE HYPERBOLA.

**255.** An hyperbola is the curve traced by a point which moves in a plane in such a manner that its distance from a given point is in a constant ratio of greater inequality to its distance from a given straight line.

**256.** The construction is the same as for the ellipse, but the position of the parts is different. As explained in § 119, *X, A'* lies on the left side of the directrix. Each directrix lies between *A* and *A'*, and the foci lie without these points. The curve consists of two branches which are open on one side. The branches lie entirely within two vertical angles formed by two straight lines passing through the center which are called the asymptotes. These are tangents to the curve at infinity.

**257.** The hyperbola can be defined thus: If a point *P* move in such a manner that $PN^2 : AN \cdot NA'$ is a constant ratio, *PN* being the distance of *P* from the line joining two fixed points *A* and *A'*, and *N* not being between *A* and *A'*, the locus of *P* is an hyperbola, of which *AA'* is the transverse axis.

This corresponds to the equation

$$y^2 = \frac{b^2}{a^2} (2ax + x^2),$$

where the origin is at the right-hand vertex of the hyperbola.

Fig. 74 shows how points on the curve may be found by the application of this formula.

Let *C* be the center and *A* the vertex of the curve.

$$CB' = CB = b;$$
$$CA' = CA = CA' = a.$$

Fold $CD$ any line through $C$ and make $CD = CA$. Fold $DN$ perpendicular to $CD$. Fold $NQ$ perpendicular to $CA$ and make $NQ = DN$. Fold $QA''$ cutting $CA$ in $S$. Fold $B'S$ cutting $QN$ in $P$.

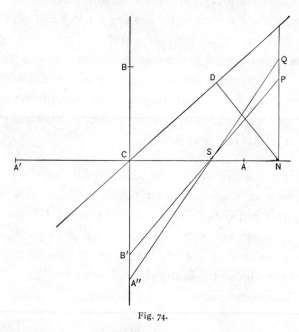

Fig. 74.

Then $P$ is a point on the curve.

For, since $DN$ is tangent to the circle on the diameter $A'A$

$$DN^2 = AN \cdot (2CA + AN),$$

or since $\qquad QN = DN,$

$$QN^2 = x(2a + x).$$

$$\frac{QN}{PN} = \frac{A''C}{B'C}.$$

Squaring, $\quad \dfrac{x(2a+x)}{y^2} = \dfrac{a^2}{b^2},$

or $\qquad\qquad y^2 = \dfrac{b^2}{a^2}(2ax + x^2).$

If $QN = b$ then $N$ is the focus and $CD$ is one of the asymptotes. If we complete the rectangle on $AC$ and $BC$ the asymptote is a diagonal of the rectangle.

**258.** The hyperbola can also be described by the property referred to in § 253.

**259.** An hyperbola is said to be equilateral when the transverse and conjugate axes are equal. Here $a = b$, and the equation becomes

$$y^2 = (2a + x)x.$$

In this case the construction is simpler as the ordinate of the hyperbola is itself the geometric mean between $AN$ and $A'N$, and is therefore equal to the tangent from $N$ to the circle described on $AA'$ as diameter.

**260.** The polar equation to the rectangular hyperbola, when the center is the origin and one of the axes the initial line, is

$$r^2 \cos 2\theta = a^2$$

or $r^2 = \dfrac{a}{\cos 2\theta} \cdot a.$

Let $OX$, $OY$ be the axes; divide the right angle $YOX$ into a number of equal parts. Let $XOA$, $AOB$

be two of the equal angles.  Fold *XB* at right angles
to *OX*.   Produce *BO* and take *OF* = *OX*.   Fold *OG*
perpendicular to *BF* and find *G* in *OG* such that *FGB*
is a right angle.   Take *OA* = *OG*.   Then *A* is a point
on the curve.

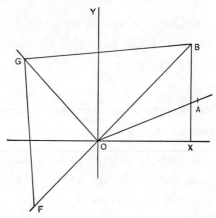

Fig. 75.

Now, the angles *XOA* and *AOB* being each $\theta$,

$$OB = \frac{a}{\cos 2\theta}.$$

And  $OA^2 = OG^2 = OB \cdot OF = \dfrac{a}{\cos 2\theta} \cdot a.$

$$\therefore \; r^2 \cos 2\theta = a^2.$$

**261.**  The points of trisection of a series of conter-
minous circular arcs lie on branches of two hyperbolas
of which the eccentricity is 2.   This theorem affords
a means of trisecting an angle.*

* See Taylor's *Ancient and Modern Geometry of Conics*, examples 308, 390
with footnote.

# XIV. MISCELLANEOUS CURVES.

**262.** I propose in this, the last chapter, to give hints for tracing certain well-known curves.

## THE CISSOID.*

**263.** This word means ivy-shaped curve. It is defined as follows: Let $OQA$ (Fig. 76) be a semicircle on the fixed diameter $OA$, and let $QM$, $RN$ be two ordinates of the semicircle equidistant from the center. Draw $OR$ cutting $QM$ in $P$. Then the locus of $P$ is the cissoid.

If $OA = 2a$, the equation to the curve is

$$y^2 (2a - x) = x^3.$$

Now, let $PR$ cut the perpendicular from $C$ in $D$ and draw $AP$ cutting $CD$ in $E$.

$$RN : CD = ON : OC = AM : AC = PM : EC,$$
$$\therefore\ RN : PM = CD : CE.$$

But $RN : PM = ON : OM = ON : AN = ON^2 : NR^2$
$$= OC^2 : CD^2,$$

$$\therefore\ CD : CE = OC^2 : CD^2.$$

If $CF$ be the geometric mean between $CD$ and $CE$,

---

*See Beman and Smith's translation of Klein's *Famous Problems of Elementary Geometry*, p. 44.

$$CD : CF = OC : CD$$

$$\therefore \quad OC : CD = CD : CF = CF : CE$$

$\therefore$ $CD$ and $CF$ are the two geometric means between $OC$ and $CE$.

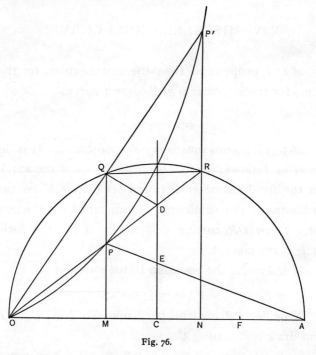

Fig. 76.

**264.** The cissoid was invented by Diocles (second century B. C.) to find two geometric means between two lines in the manner described above. $OC$ and $CE$ being given, the point $P$ was determined by the aid of the curve, and hence the point $D$.

**265.** If $PD$ and $DR$ are each equal to $OQ$, then the angle $AOQ$ is trisected by $OP$.

Draw $QR$. Then $QR$ is parallel to $OA$, and

$$DQ = DP = DR = OQ$$

$$\therefore \angle ROQ = \angle QDO = 2 \angle QRO = 2 \angle AOR.$$

### THE CONCHOID OR MUSSEL-SHAPED CURVE.*

**266.** This curve was invented by Nicomedes (c.
150 B. C.). Let $O$ be a
fixed point, $a$ its dis-
tance from a fixed line,
$DM$, and let a pencil of
rays through $O$ cut $DM$.
On each of these rays
lay off, each way from its
intersection with $DM$, a
segment $b$. The locus
of the points thus deter-
mined is the conchoid.
According as $b >$, $=$,
or $< a$, the origin is a
node, a cusp, or a con-
jugate point. The fig-
ure† represents the case
when $b > a$.

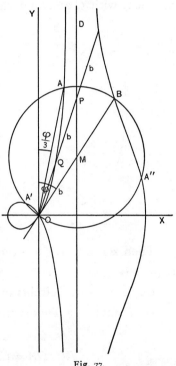

**267.** This curve also
was employed for finding

Fig. 77.

two geometric means, and for the trisection of an angle.

*See Beman and Smith's translation of Klein's *Famous Problems of Ele-
mentary Geometry*, p. 40.

†From Beman and Smith's translation of Klein's *Famous Problems of
Elementary Geometry*, p. 46.

Let $OA$ be the longer of the two lines of which two geometric means are required.

Bisect $OA$ in $B$; with $O$ as a center and $OB$ as a radius describe a circle. Place a chord $BC$ in the circle equal to the shorter of the given lines. Draw $AC$ and produce $AC$ and $BC$ to $D$ and $E$, two points collinear with $O$ and such that $DE = OB$, or $BA$.

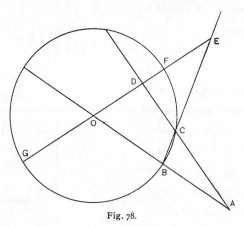

Fig. 78.

Then $ED$ and $CE$ are the two mean proportionals required.

Let $OE$ cut the circles in $F$ and $G$.

By Menelaus's Theorem,*

$$BC \cdot ED \cdot OA = CE \cdot OD \cdot BA$$
$$\therefore \ BC \cdot OA = CE \cdot OD$$
$$\text{or } \frac{BC}{CE} = \frac{OD}{OA}$$
$$\therefore \ \frac{BE}{CE} = \frac{OD + OA}{OA} = \frac{GE}{OA}.$$

*See Beman and Smith's *New Plane and Solid Geometry*, p. 240.

But $\qquad GE \cdot EF = BE \cdot EC.$

$\qquad \therefore\ GE \cdot OD = BE \cdot EC.$

$\qquad \therefore\ OA \cdot OD = EC^2.$

$\qquad \therefore\ OA : CE = CE : OD = OD : BC.$

The position of $E$ is found by the aid of the conchoid of which $AD$ is the asymptote, $O$ the focus, and $DE$ the constant intercept.

**268.** The trisection of the angle is thus effected. In Fig. 77, let $\phi = \angle MOY$, the angle to be trisected. On $OM$ lay off $OM = b$, any arbitrary length. With $M$ as a center and a radius $b$ describe a circle, and through $M$ perpendicular to the axis of $X$ with origin $O$ draw a vertical line representing the asymptote of the conchoid to be constructed. Construct the conchoid. Connect $O$ with $A$, the intersection of the circle and the conchoid. Then is $\angle AOY$ one third of $\varphi$.*

### THE WITCH.

**269.** If $OQA$ (Fig. 79) be a semicircle and $NQ$ an ordinate of it, and $NP$ be taken a fourth proportional to $ON$, $OA$ and $QN$, then the locus of $P$ is the witch.

Fold $AM$ at right angles to $OA$.

Fold through $O$, $Q$. and $M$.

Complete the rectangle $NAMP$.

$$PN : QN = OM : OQ$$
$$= OA : ON.$$

* Beman and Smith's translation of Klein's *Famous Problems of Elementary Geometry*, p. 46.

Therefore $P$ is a point on the curve.

Its equation is,

$$xy^2 = a^2 (a - x).$$

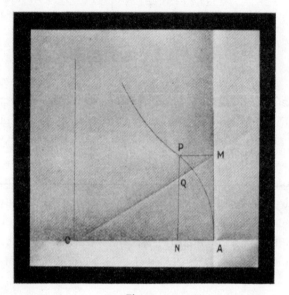

Fig. 79.

This curve was proposed by a lady, Maria Gaetana Agnesi, Professor of Mathematics at Bologna.

### THE CUBICAL PARABOLA.

**270.** The equation to this curve is $a^2y = x^3$.

Let $OX$ and $OY$ be the rectangular axes, $OA = a$, and $OX = x$.

In the axis $OY$ take $OB = x$.

Draw $BA$ and draw $AC$ at right angles to $AB$ cutting the axis $OY$ in $C$.

Draw *CX*, and draw *XY* at right angles to *CX*.

Complete the rectangle *XOY*.

*P* is a point on the curve.

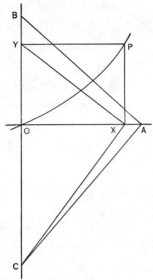

Fig. 80.

$$y = XP = OY = \frac{x^2}{OC} = x^2 \cdot \frac{x}{a^2} = \frac{x^3}{a^2}$$

$$\therefore \ a^2 y = x^3.$$

## THE HARMONIC CURVE OR CURVE OF SINES.

**271.** This is the curve in which a musical string vibrates when sounded. The ordinates are proportional to the sines of angles which are the same fractions of four right angles that the corresponding abscissas are of some given length.

Let *AB* (Fig. 81) be the given length. Produce *BA*

to $C$ and fold $AD$ perpendicular to $AB$. Divide the right angle $DAC$ into a number of equal parts, say, four. Mark on each radius a length equal to the amplitude of the vibration, $AC = AP = AQ = AR = AD$.

From points $P$, $Q$, $R$ fold perpendiculars to $AC$; then $PP'$, $QQ'$, $RR'$, and $DA$ are proportional to the sines of the angles $PAC$, $QAC$, $RAC$, $DAC$.

Now, bisect $AB$ in $E$ and divide $AE$ and $EB$ into twice the number of equal parts chosen for the right

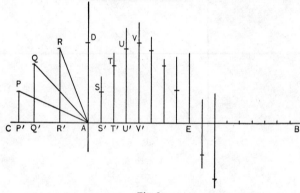

Fig. 81.

angle. Draw the successive ordinates $SS'$, $TT'$, $UU'$, $VV'$, etc., equal to $PP'$, $QQ'$, $RR'$, $DA$, etc. Then $S$, $T$, $U$, $V$ are points on the curve, and $V$ is the highest point on it. By folding on $VV'$ and pricking through $S$, $T$, $U$, $V$, we get corresponding points on the portion of the curve $VE$. The portion of the curve corresponding to $EB$ is equal to $AVE$ but lies on the opposite side of $AB$. The length from $A$ to $F$ is half a wave length, which will be repeated from $E$

to $B$ on the other side of $AB$.  $E$ is a point of inflection on the curve, the radius of curvature there becoming infinite.

### THE OVALS OF CASSINI.

**272.** When a point moves in a plane so that the product of its distances from two fixed points in the plane is constant, it traces out one of Cassini's ovals. The fixed points are called the foci.  The equation of

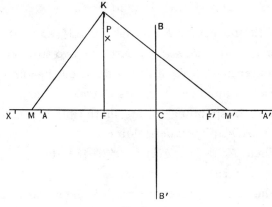

Fig. 82.

the curve is $rr' = k^2$, where $r$ and $r'$ are the distances of any point on the curve from the foci and $k$ is a constant.

Let $F$ and $F'$ be the foci.  Fold through $F$ and $F'$.  Bisect $FF'$ in $C$, and fold $BCB'$ perpendicular to $FF'$.  Find points $B$ and $B'$ such that $FB$ and $FB'$ are each $= k$.  Then $B$ and $B'$ are evidently points on the curve.

Fold $FK$ perpendicular to $FF'$ and make $FK = k$, and on $FF'$ take $CA$ and $CA'$ each equal to $CK$. Then $A$ and $A'$ are points on the curve.

For
$$CA^2 = CK^2 = CF^2 + FK^2.$$
$$\therefore CA^2 - CF^2 = k^2 = (CA + CF)(CA - CF)$$
$$= F'A \cdot FA.$$

Produce $FA$ and take $AT = FK$. In $AT$ take a point $M$ and draw $MK$. Fold $KM'$ perpendicular to $MK$ meeting $FA'$ in $M'$.

Then $FM \cdot FM' = k^2$.

With the center $F$ and radius $FM$, and with the center $F'$ and radius $FM'$, describe two arcs cutting each other in $P$. Then $P$ is a point on the curve.

When a number of points between $A$ and $B$ are found, corresponding points in the other quadrants can be marked by paper folding.

When $FF' = \sqrt{2}\,k$ and $rr' = \frac{1}{2}k^2$ the curve assumes the form of a lemniscate. (§ 279.)

When $FF'$ is greater than $\sqrt{2}\,k$, the curve consists of two distinct ovals, one about each focus.

### THE LOGARITHMIC CURVE.

**273.** The equation to this curve is $y = a^x$.

The ordinate at the origin is unity.

If the abscissa increases arithmetically, the ordinate increases geometrically.

The values of $y$ for integral values of $x$ can be obtained by the process given in § 108.

The curve extends to infinity in the angular space *XOY*.

If $x$ be negative $y = \dfrac{1}{a^x}$ and approaches zero as $x$ increases numerically. The negative side of the axis *OX* is therefore an asymptote to the curve.

### THE COMMON CATENARY.

**274.** The catenary is the form assumed by a heavy inextensible string freely suspended from two points and hanging under the action of gravity.

The equation of the curve is

$$y = \frac{c}{2}\left(e^{\frac{x}{c}} + e^{-\frac{x}{c}}\right)$$

the axis of $y$ being a vertical line through the lowest point of the curve, and the axis of $x$ a horizontal line in the plane of the string at a distance $c$ below the lowest point; $c$ is the parameter of the curve, and $e$ the base of the natural system of logarithms.

$$\text{When } x = c, \quad y = \frac{c}{2}(e^1 + e^{-1})$$

$$\text{when } x = 2c, \quad y = \frac{c}{2}(e^2 + e^{-2}) \text{ and so on.}$$

**275.** From the equation

$$y = \frac{c}{2}\left(e^{\frac{1}{2}} + e^{-\frac{1}{2}}\right)$$

$e$ can be determined graphically.

$$ce - 2y\sqrt{e} + c = 0$$

$$\sqrt{e} = \frac{1}{c}\left(y + \sqrt{y^2 - c^2}\right)$$

$$c\sqrt{e} = y + \sqrt{y^2 - c^2}.$$

$\sqrt{y^2 - c^2}$ is found by taking the geometric mean between $y + c$ and $y - c$.

### THE CARDIOID OR HEART-SHAPED CURVE.

**276.** From a fixed point $O$ on a circle of radius $a$ draw a pencil of lines and take off on each ray, measured both ways from the circumference, a segment equal to $2a$. The ends of these lines lie on a cardioid.

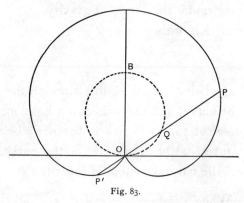

Fig. 83.

The equation to the curve is $r = a(1 + \cos\theta)$.

The origin is a cusp on the curve. The cardioid is the inverse of the parabola with reference to its focus as center of inversion.

### THE LIMACON.

**277.** From a fixed point on a circle, draw a number of chords, and take off a constant length on each of these lines measured both ways from the circumference of the circle.

If the constant length is equal to the diameter of the circle, the curve is a cardioid.

If it be greater than the diameter, the curve is altogether outside the circle.

If it be less than the diameter, a portion of the curve lies inside the circle in the form of a loop.

If the constant length is exactly half the diameter, the curve is called the trisectrix, since by its aid any angle can be trisected.

The equation is $r = a\cos\theta + b$.

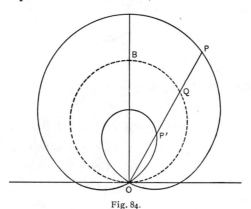

Fig. 84.

The first sort of limaçon is the inverse of an ellipse; and the second sort is the inverse of an hyperbola, with reference to a focus as a center. The loop is the inverse of the branch about the other focus.

**278.** The trisectrix is applied as follows:

Let $AOB$ be the given angle. Take $OA$, $OB$ equal to the radius of the circle. Describe a circle with the center $O$ and radius $OA$ or $OB$. Produce $AO$ indefinitely beyond the circle. Apply the trisectrix so that $O$ may correspond to the center of the circle and $OB$ the axis of the loop. Let the outer curve cut $AO$ produced in $C$. Draw $BC$ cutting the circle in $D$, Draw $OD$.

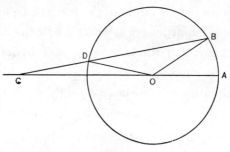

Fig. 85.

Then $\angle ACB$ is $\frac{1}{3}$ of $\angle AOB$.

For $CD = DO = OB$.

$\therefore \quad \angle AOB = \angle ACB + \angle CBO$
$= \angle ACB + \angle ODB$
$= \angle ACB + 2\angle ACB$
$= 3\angle ACB.$

### THE LEMNISCATE OF BERNOULLI.

**279.** The polar equation to the curve is
$$r^2 = a^2 \cos 2\theta.$$
Let $O$ be the origin, and $OA = a$.

Produce $AO$, and draw $OD$ at right angles to $OA$

Take the angle $AOP = \theta$ and $AOB = 2\theta$.

Draw $AB$ perpendicular to $OB$.

In $AO$ produced take $OC = OB$.

Find $D$ in $OD$ such that $CDA$ is a right angle.

Take $OP = OD$.

$P$ is a point on the curve.

$$r^2 = OD^2 = OC \cdot OA$$
$$= OB \cdot OA$$
$$= a \cos 2\theta \cdot a$$
$$= a^2 \cos 2\theta.$$

As stated above, this curve is a particular case of the ovals of Cassini.

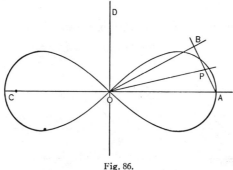

Fig. 86.

It is the inverse of the rectangular hyperbola, with reference to its center as center of inversion, and also its pedal with respect to the center.

The area of the curve is $a^2$.

### THE CYCLOID.

**280.** The cycloid is the path described by a point on the circumference of a circle which is supposed to roll upon a fixed straight line.

Let $A$ and $A'$ be the positions of the generating point when in contact with the fixed line after one

complete revolution of the circle. Then $AA'$ is equal
to the circumference of the circle.

The circumference of a circle may be obtained in
length in this way. Wrap a strip of paper round a
circular object, e. g., the cylinder in Kindergarten
gift No. II., and mark off two coincident points. Un-
fold the paper and fold through the points. Then the
straight line between the two points is equal to the
circumference corresponding to the diameter of the
cylinder.

By proportion, the circumference corresponding
to any diameter can be found and *vice versa*.

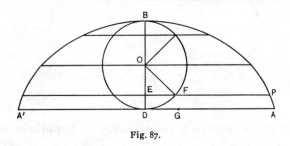

Fig. 87.

Bisect $AA'$ in $D$ and draw $DB$ at right angles to
$AA'$, and equal to the diameter of the generating
circle.

Then $A$, $A'$ and $B$ are points on the curve.

Find $O$ the middle point of $BD$.

Fold a number of radii of the generating circle
through $O$ dividing the semi-circumference to the
right into equal arcs, say, four.

Divide $AD$ into the same number of equal parts.

Through the ends of the diameters fold lines at right angles to *BD*.

Let *EFP* be one of these lines, *F* being the end of a radius, and let *G* be the corresponding point of section of *AD*, commencing from *D*. Mark off *FP* equal to *GA* or to the length of arc *BF*.

Then *P* is a point on the curve.

Other points corresponding to other points of section of *AD* may be marked in the same way.

The curve is symmetric to the axis *BD* and corresponding points on the other half of the curve can be marked by folding on *BD*.

The length of the curve is 4 times *BD* and its area 3 times the area of the generating circle.

### THE TROCHOID.

**281.** If as in the cycloid, a circle rolls along a straight line, any point in the plane of the circle but not on its circumference traces out the curve called a trochoid.

### THE EPICYCLOID.

**282.** An epicycloid is the path described by a point on the circumference of a circle which rolls on the circumference of another fixed circle touching it on the outside.

### THE HYPOCYCLOID.

**283.** If the rolling circle touches the inside of the fixed circle, the curve traced by a point on the circumference of the former is a hypocycloid.

When the radius of the rolling circle is a sub-multiple of the fixed circle, the circumference of the latter has to be divided in the same ratio.

These sections being divided into a number of equal parts, the position of the center of the rolling circle and of the generating point corresponding to each point of section of the fixed circle can be found by dividing the circumference of the rolling circle into the same number of equal parts.

### THE QUADRATRIX.*

**284.** Let $OACB$ be a square. If the radius $OA$ of a circle rotate uniformly round the center $O$ from the position $OA$ through a right angle to $OB$ and if in the same time a straight line drawn perpendicular to $OB$ move uniformly parallel to itself from the position $OA$ to $BC$; the locus of their intersection will be the quadratrix.

This curve was invented by Hippias of Elis (420 B. C.) for the multisection of an angle.

If $P$ and $P'$ are points on the curve, the angles $AOP$ and $AOP'$ are to one another as the ordinates of the respective points.

### THE SPIRAL OF ARCHIMEDES.

**285.** If the line $OA$ revolve uniformly round $O$ as center, while point $P$ moves uniformly from $O$ along $OA$, then the point $P$ will describe the spiral of Archimedes.

---

*Beman and Smith's translation of Klein's *Famous Problems of Elementary Geometry*, p. 57.

# A CATALOGUE OF SELECTED DOVER BOOKS
## IN ALL FIELDS OF INTEREST

# A CATALOGUE OF SELECTED DOVER BOOKS
## IN ALL FIELDS OF INTEREST

LEATHER TOOLING AND CARVING, Chris H. Groneman. One of few books concentrating on tooling and carving, with complete instructions and grid designs for 39 projects ranging from bookmarks to bags. 148 illustrations. 111pp. 7⅞ x 10.
23061-9 Pa. $2.50

THE CODEX NUTTALL, A PICTURE MANUSCRIPT FROM ANCIENT MEXICO, as first edited by Zelia Nuttall. Only inexpensive edition, in full color, of a pre-Columbian Mexican (Mixtec) book. 88 color plates show kings, gods, heroes, temples, sacrifices. New explanatory, historical introduction by Arthur G. Miller. 96pp. 11⅜ x 8½.
23168-2 Pa. $7.50

AMERICAN PRIMITIVE PAINTING, Jean Lipman. Classic collection of an enduring American tradition. 109 plates, 8 in full color—portraits, landscapes, Biblical and historical scenes, etc., showing family groups, farm life, and so on. 80pp. of lucid text. 8⅜ x 11¼.
22815-0 Pa. $5.00

WILL BRADLEY: HIS GRAPHIC ART, edited by Clarence P. Hornung. Striking collection of work by foremost practitioner of Art Nouveau in America: posters, cover designs, sample pages, advertisements, other illustrations. 97 plates, including 8 in full color and 19 in two colors. 97pp. 9⅜ x 12¼.
20701-3 Pa. $4.00
22120-2 Clothbd. $10.00

AN ATLAS OF ANATOMY FOR ARTISTS, Fritz Schider. Finest text, working book. Full text, plus anatomical illustrations; plates by great artists showing anatomy. 593 illustrations. 192pp. 7⅞ x 10¾.
20241-0 Clothbd. $6.95

THE GIBSON GIRL AND HER AMERICA, Charles Dana Gibson. 155 finest drawings of effervescent world of 1900-1910: the Gibson Girl and her loves, amusements, adventures, Mr. Pipp, etc. Selected by E. Gillon; introduction by Henry Pitz. 144pp. 8¼ x 11⅜.
21986-0 Pa. $3.50

STAINED GLASS CRAFT, J.A.F. Divine, G. Blachford. One of the very few books that tell the beginner exactly what he needs to know: planning cuts, making shapes, avoiding design weaknesses, fitting glass, etc. 93 illustrations. 115pp.
22812-6 Pa. $1.75

CREATIVE LITHOGRAPHY AND HOW TO DO IT, Grant Arnold. Lithography as art form: working directly on stone, transfer of drawings, lithotint, mezzotint, color printing; also metal plates. Detailed, thorough. 27 illustrations. 214pp.
21208-4 Pa. $3.50

DESIGN MOTIFS OF ANCIENT MEXICO, Jorge Enciso. Vigorous, powerful ceramic stamp impressions — Maya, Aztec, Toltec, Olmec. Serpents, gods, priests, dancers, etc. 153pp. 6⅛ x 9¼. 20084-1 Pa. $2.50

AMERICAN INDIAN DESIGN AND DECORATION, Leroy Appleton. Full text, plus more than 700 precise drawings of Inca, Maya, Aztec, Pueblo, Plains, NW Coast basketry, sculpture, painting, pottery, sand paintings, metal, etc. 4 plates in color. 279pp. 8⅜ x 11¼. 22704-9 Pa.$5.00

CHINESE LATTICE DESIGNS, Daniel S. Dye. Incredibly beautiful geometric designs: circles, voluted, simple dissections, etc. Inexhaustible source of ideas, motifs. 1239 illustrations. 469pp. 6⅛ x 9¼. 23096-1 Pa. $5.00

JAPANESE DESIGN MOTIFS, Matsuya Co. Mon, or heraldic designs. Over 4000 typical, beautiful designs: birds, animals, flowers, swords, fans, geometric; all beautifully stylized. 213pp. 11⅜ x 8¼. 22874-6 Pa. $5.00

PERSPECTIVE, Jan Vredeman de Vries. 73 perspective plates from 1604 edition; buildings, townscapes, stairways, fantastic scenes. Remarkable for beauty, surrealistic atmosphere; real eye-catchers. Introduction by Adolf Placzek. 74pp. 11⅜ x 8¼. 20186-4 Pa. $2.75

EARLY AMERICAN DESIGN MOTIFS. Suzanne E. Chapman. 497 motifs, designs, from painting on wood, ceramics, appliqué, glassware, samplers, metal work, etc. Florals, landscapes, birds and animals, geometrics, letters, etc. Inexhaustible. Enlarged edition. 138pp. 8⅜ x 11¼. 22985-8 Pa. $3.50
23084-8 Clothbd. $7.95

VICTORIAN STENCILS FOR DESIGN AND DECORATION, edited by E.V. Gillon, Jr. 113 wonderful ornate Victorian pieces from German sources; florals, geometrics; borders, corner pieces; bird motifs, etc. 64pp. 9⅜ x 12¼. 21995-X Pa. $3.00

ART NOUVEAU: AN ANTHOLOGY OF DESIGN AND ILLUSTRATION FROM THE STUDIO, edited by E.V. Gillon, Jr. Graphic arts: book jackets, posters, engravings, illustrations, decorations; Crane, Beardsley, Bradley and many others. Inexhaustible. 92pp. 8⅛ x 11. 22388-4 Pa. $2.50

ORIGINAL ART DECO DESIGNS, William Rowe. First-rate, highly imaginative modern Art Deco frames, borders, compositions, alphabets, florals, insectals, Wurlitzer-types, etc. Much finest modern Art Deco. 80 plates, 8 in color. 8⅜ x 11¼. 22567-4 Pa. $3.50

HANDBOOK OF DESIGNS AND DEVICES, Clarence P. Hornung. Over 1800 basic geometric designs based on circle, triangle, square, scroll, cross, etc. Largest such collection in existence. 261pp. 20125-2 Pa. $2.75

150 MASTERPIECES OF DRAWING, edited by Anthony Toney. 150 plates, early 15th century to end of 18th century; Rembrandt, Michelangelo, Dürer, Fragonard, Watteau, Wouwerman, many others. 150pp. 8⅜ x 11¼.     21032-4 Pa. $4.00

THE GOLDEN AGE OF THE POSTER, Hayward and Blanche Cirker. 70 extraordinary posters in full colors, from Maîtres de l'Affiche, Mucha, Lautrec, Bradley, Cheret, Beardsley, many others. 9⅜ x 12¼.     22753-7 Pa. $5.95
21718-3 Clothbd. $7.95

SIMPLICISSIMUS, selection, translations and text by Stanley Appelbaum. 180 satirical drawings, 16 in full color, from the famous German weekly magazine in the years 1896 to 1926. 24 artists included: Grosz, Kley, Pascin, Kubin, Kollwitz, plus Heine, Thöny, Bruno Paul, others. 172pp. 8½ x 12¼.     23098-8 Pa. $5.00
23099-6 Clothbd. $10.00

THE EARLY WORK OF AUBREY BEARDSLEY, Aubrey Beardsley. 157 plates, 2 in color: Manon Lescaut, Madame Bovary, Morte d'Arthur, Salome, other. Introduction by H. Marillier. 175pp. 8½ x 11.     21816-3 Pa. $4.00

THE LATER WORK OF AUBREY BEARDSLEY, Aubrey Beardsley. Exotic masterpieces of full maturity: Venus and Tannhäuser, Lysistrata, Rape of the Lock, Volpone, Savoy material, etc. 174 plates, 2 in color. 176pp. 8½ x 11.     21817-1 Pa. $4.00

DRAWINGS OF WILLIAM BLAKE, William Blake. 92 plates from Book of Job, Divine Comedy, Paradise Lost, visionary heads, mythological figures, Laocoön, etc. Selection, introduction, commentary by Sir Geoffrey Keynes. 178pp. 8½ x 11.     22303-5 Pa. $4.00

LONDON: A PILGRIMAGE, Gustave Doré, Blanchard Jerrold. Squalor, riches, misery, beauty of mid-Victorian metropolis; 55 wonderful plates, 125 other illustrations, full social, cultural text by Jerrold. 191pp. of text. 8⅛ x 11.     22306-X Pa. $6.00

THE COMPLETE WOODCUTS OF ALBRECHT DÜRER, edited by Dr. W. Kurth. 346 in all: Old Testament, St. Jerome, Passion, Life of Virgin, Apocalypse, many others. Introduction by Campbell Dodgson. 285pp. 8½ x 12¼.     21097-9 Pa. $6.00

THE DISASTERS OF WAR, Francisco Goya. 83 etchings record horrors of Napoleonic wars in Spain and war in general. Reprint of 1st edition, plus 3 additional plates. Introduction by Philip Hofer. 97pp. 9⅜ x 8¼.     21872-4 Pa. $3.50

ENGRAVINGS OF HOGARTH, William Hogarth. 101 of Hogarth's greatest works: Rake's Progress, Harlot's Progress, Illustrations for Hudibras, Midnight Modern Conversation, Before and After, Beer Street and Gin Lane, many more. Full commentary. 256pp. 11 x 14.     22479-1 Pa. $7.95,

PRIMITIVE ART, Franz Boas. Great anthropologist on ceramics, textiles, wood, stone, metal, etc.; patterns, technology, symbols, styles. All areas, but fullest on Northwest Coast Indians. 350 illustrations. 378pp.     20025-6 Pa. $3.75

MOTHER GOOSE'S MELODIES. Facsimile of fabulously rare Munroe and Francis "copyright 1833" Boston edition. Familiar and unusual rhymes, wonderful old woodcut illustrations. Edited by E.F. Bleiler. 128pp. 4½ x 6⅜. 22577-1 Pa. $1.50

MOTHER GOOSE IN HIEROGLYPHICS. Favorite nursery rhymes presented in rebus form for children. Fascinating 1849 edition reproduced in toto, with key. Introduction by E.F. Bleiler. About 400 woodcuts. 64pp. 6⅞ x 5¼. 20745-5 Pa. $1.50

PETER PIPER'S PRACTICAL PRINCIPLES OF PLAIN & PERFECT PRONUNCIATION. Alliterative jingles and tongue-twisters. Reproduction in full of 1830 first American edition. 25 spirited woodcuts. 32pp. 4½ x 6⅜. 22560-7 Pa. $1.25

THE NIGHT BEFORE CHRISTMAS, Clement Moore. Full text, and woodcuts from original 1848 book. Also critical, historical material. 19 illustrations. 40pp. 4⅝ x 6. 22797-9 Pa. $1.35

THE KING OF THE GOLDEN RIVER, John Ruskin. Victorian children's classic of three brothers, their attempts to reach the Golden River, what becomes of them. Facsimile of original 1889 edition. 22 illustrations. 56pp. 4⅝ x 6⅜. 20066-3 Pa. $1.50

DREAMS OF THE RAREBIT FIEND, Winsor McCay. Pioneer cartoon strip, unexcelled for beauty, imagination, in 60 full sequences. Incredible technical virtuosity, wonderful visual wit. Historical introduction. 62pp. 8⅜ x 11¼. 21347-1 Pa. $2.50

THE KATZENJAMMER KIDS, Rudolf Dirks. In full color, 14 strips from 1906-7; full of imagination, characteristic humor. Classic of great historical importance. Introduction by August Derleth. 32pp. 9¼ x 12¼. 23005-8 Pa. $2.00

LITTLE ORPHAN ANNIE AND LITTLE ORPHAN ANNIE IN COSMIC CITY, Harold Gray. Two great sequences from the early strips: our curly-haired heroine defends the Warbucks' financial empire and, then, takes on meanie Phineas P. Pinchpenny. Leapin' lizards! 178pp. 6⅛ x 8⅜. 23107-0 Pa. $2.00

WHEN A FELLER NEEDS A FRIEND, Clare Briggs. 122 cartoons by one of the greatest newspaper cartoonists of the early 20th century — about growing up, making a living, family life, daily frustrations and occasional triumphs. 121pp. 8½ x 9½. 23148-8 Pa. $2.50

ABSOLUTELY MAD INVENTIONS, A.E. Brown, H.A. Jeffcott. Hilarious, useless, or merely absurd inventions all granted patents by the U.S. Patent Office. Edible tie pin, mechanical hat tipper, etc. 57 illustrations. 125pp. 22596-8 Pa. $1.50

THE DEVIL'S DICTIONARY, Ambrose Bierce. Barbed, bitter, brilliant witticisms in the form of a dictionary. Best, most ferocious satire America has produced. 145pp. 20487-1 Pa. $1.75

# CATALOGUE OF DOVER BOOKS

THE BEST DR. THORNDYKE DETECTIVE STORIES, R. Austin Freeman. The Case of Oscar Brodski, The Moabite Cipher, and 5 other favorites featuring the great scientific detective, plus his long-believed-lost first adventure — 31 New Inn — reprinted here for the first time. Edited by E.F. Bleiler. USO 20388-3 Pa. $3.00

BEST "THINKING MACHINE" DETECTIVE STORIES, Jacques Futrelle. The Problem of Cell 13 and 11 other stories about Prof. Augustus S.F.X. Van Dusen, including two "lost" stories. First reprinting of several. Edited by E.F. Bleiler. 241pp.
20537-1 Pa. $3.00

UNCLE SILAS, J. Sheridan LeFanu. Victorian Gothic mystery novel, considered by many best of period, even better than Collins or Dickens. Wonderful psychological terror. Introduction by Frederick Shroyer. 436pp. 21715-9 Pa. $4.00

BEST DR. POGGIOLI DETECTIVE STORIES, T.S. Stribling. 15 best stories from EQMM and The Saint offer new adventures in Mexico, Florida, Tennessee hills as Poggioli unravels mysteries and combats Count Jalacki. 217pp. 23227-1 Pa. $3.00

EIGHT DIME NOVELS, selected with an introduction by E.F. Bleiler. Adventures of Old King Brady, Frank James, Nick Carter, Deadwood Dick, Buffalo Bill, The Steam Man, Frank Merriwell, and Horatio Alger — 1877 to 1905. Important, entertaining popular literature in facsimile reprint, with original covers. 190pp. 9 x 12. 22975-0 Pa. $3.50

ALICE'S ADVENTURES UNDER GROUND, Lewis Carroll. Facsimile of ms. Carroll gave Alice Liddell in 1864. Different in many ways from final Alice. Handlettered, illustrated by Carroll. Introduction by Martin Gardner. 128pp. 21482-6 Pa. $2.00

ALICE IN WONDERLAND COLORING BOOK, Lewis Carroll. Pictures by John Tenniel. Large-size versions of the famous illustrations of Alice, Cheshire Cat, Mad Hatter and all the others, waiting for your crayons. Abridged text. 36 illustrations. 64pp. 8¼ x 11. 22853-3 Pa. $1.50

AVENTURES D'ALICE AU PAYS DES MERVEILLES, Lewis Carroll. Bué's translation of "Alice" into French, supervised by Carroll himself. Novel way to learn language. (No English text.) 42 Tenniel illustrations. 196pp. 22836-3 Pa. $3.00

MYTHS AND FOLK TALES OF IRELAND, Jeremiah Curtin. 11 stories that are Irish versions of European fairy tales and 9 stories from the Fenian cycle — 20 tales of legend and magic that comprise an essential work in the history of folklore. 256pp. 22430-9 Pa. $3.00

EAST O' THE SUN AND WEST O' THE MOON, George W. Dasent. Only full edition of favorite, wonderful Norwegian fairytales — Why the Sea is Salt, Boots and the Troll, etc. — with 77 illustrations by Kittelsen & Werenskiöld. 418pp.
22521-6 Pa. $4.50

PERRAULT'S FAIRY TALES, Charles Perrault and Gustave Doré. Original versions of Cinderella, Sleeping Beauty, Little Red Riding Hood, etc. in best translation, with 34 wonderful illustrations by Gustave Doré. 117pp. 8⅛ x 11. 22311-6 Pa. $2.50

EARLY NEW ENGLAND GRAVESTONE RUBBINGS, Edmund V. Gillon, Jr. 43 photographs, 226 rubbings show heavily symbolic, macabre, sometimes humorous primitive American art. Up to early 19th century. 207pp. 8⅜ x 11¼.
21380-3 Pa. $4.00

L.J.M. DAGUERRE: THE HISTORY OF THE DIORAMA AND THE DAGUERREOTYPE, Helmut and Alison Gernsheim. Definitive account. Early history, life and work of Daguerre; discovery of daguerreotype process; diffusion abroad; other early photography. 124 illustrations. 226pp. 6⅙ x 9¼.
22290-X Pa. $4.00

PHOTOGRAPHY AND THE AMERICAN SCENE, Robert Taft. The basic book on American photography as art, recording form, 1839-1889. Development, influence on society, great photographers, types (portraits, war, frontier, etc.), whatever else needed. Inexhaustible. Illustrated with 322 early photos, daguerreotypes, tintypes, stereo slides, etc. 546pp. 6⅛ x 9¼.
21201-7 Pa. $5.95

PHOTOGRAPHIC SKETCHBOOK OF THE CIVIL WAR, Alexander Gardner. Reproduction of 1866 volume with 100 on-the-field photographs: Manassas, Lincoln on battlefield, slave pens, etc. Introduction by E.F. Bleiler. 224pp. 10¾ x 9.
22731-6 Pa. $6.00

THE MOVIES: A PICTURE QUIZ BOOK, Stanley Appelbaum & Hayward Cirker. Match stars with their movies, name actors and actresses, test your movie skill with 241 stills from 236 great movies, 1902-1959. Indexes of performers and films. 128pp. 8⅜ x 9¼.
20222-4 Pa. $2.50

THE TALKIES, Richard Griffith. Anthology of features, articles from Photoplay, 1928-1940, reproduced complete. Stars, famous movies, technical features, fabulous ads, etc.; Garbo, Chaplin, King Kong, Lubitsch, etc. 4 color plates, scores of illustrations. 327pp. 8⅜ x 11¼.
22762-6 Pa. $6.95

THE MOVIE MUSICAL FROM VITAPHONE TO "42ND STREET," edited by Miles Kreuger. Relive the rise of the movie musical as reported in the pages of Photoplay magazine (1926-1933): every movie review, cast list, ad, and record review; every significant feature article, production still, biography, forecast, and gossip story. Profusely illustrated. 367pp. 8⅜ x 11¼.
23154-2 Pa. $7.95

JOHANN SEBASTIAN BACH, Philipp Spitta. Great classic of biography, musical commentary, with hundreds of pieces analyzed. Also good for Bach's contemporaries. 450 musical examples. Total of 1799pp.
EUK 22278-0, 22279-9 Clothbd., Two vol. set $25.00

BEETHOVEN AND HIS NINE SYMPHONIES, Sir George Grove. Thorough history, analysis, commentary on symphonies and some related pieces. For either beginner or advanced student. 436 musical passages. 407pp.
20334-4 Pa. $4.00

MOZART AND HIS PIANO CONCERTOS, Cuthbert Girdlestone. The only full-length study. Detailed analyses of all 21 concertos, sources; 417 musical examples. 509pp.
21271-8 Pa. $6.00

THE FITZWILLIAM VIRGINAL BOOK, edited by J. Fuller Maitland, W.B. Squire. Famous early 17th century collection of keyboard music, 300 works by Morley, Byrd, Bull, Gibbons, etc. Modern notation. Total of 938pp. 8⅜ x 11.
ECE 21068-5, 21069-3 Pa., Two vol. set $15.00

COMPLETE STRING QUARTETS, Wolfgang A. Mozart. Breitkopf and Härtel edition. All 23 string quartets plus alternate slow movement to K156. Study score. 277pp. 9⅜ x 12¼.
22372-8 Pa. $6.00

COMPLETE SONG CYCLES, Franz Schubert. Complete piano, vocal music of Die Schöne Müllerin, Die Winterreise, Schwanengesang. Also Drinker English singing translations. Breitkopf and Härtel edition. 217pp. 9⅜ x 12¼.
22649-2 Pa. $5.00

THE COMPLETE PRELUDES AND ETUDES FOR PIANOFORTE SOLO, Alexander Scriabin. All the preludes and etudes including many perfectly spun miniatures. Edited by K.N. Igumnov and Y.I. Mil'shteyn. 250pp. 9 x 12.
22919-X Pa. $6.00

TRISTAN UND ISOLDE, Richard Wagner. Full orchestral score with complete instrumentation. Do not confuse with piano reduction. Commentary by Felix Mottl, great Wagnerian conductor and scholar. Study score. 655pp. 8⅛ x 11.
22915-7 Pa. $11.95

FAVORITE SONGS OF THE NINETIES, ed. Robert Fremont. Full reproduction, including covers, of 88 favorites: Ta-Ra-Ra-Boom-De-Aye, The Band Played On, Bird in a Gilded Cage, Under the Bamboo Tree, After the Ball, etc. 401pp. 9 x 12.
EBE 21536-9 Pa. $6.95

SOUSA'S GREAT MARCHES IN PIANO TRANSCRIPTION: ORIGINAL SHEET MUSIC OF 23 WORKS, John Philip Sousa. Selected by Lester S. Levy. Playing edition includes: The Stars and Stripes Forever, The Thunderer, The Gladiator, King Cotton, Washington Post, much more. 24 illustrations. 111pp. 9 x 12.
USO 23132-1 Pa. $3.50

CLASSIC PIANO RAGS, selected with an introduction by Rudi Blesh. Best ragtime music (1897-1922) by Scott Joplin, James Scott, Joseph F. Lamb, Tom Turpin, 9 others. Printed from best original sheet music, plus covers. 364pp. 9 x 12.
EBE 20469-3 Pa. $7.50

ANALYSIS OF CHINESE CHARACTERS, C.D. Wilder, J.H. Ingram. 1000 most important characters analyzed according to primitives, phonetics, historical development. Traditional method offers mnemonic aid to beginner, intermediate student of Chinese, Japanese. 365pp.
23045-7 Pa. $4.00

MODERN CHINESE: A BASIC COURSE, Faculty of Peking University. Self study, classroom course in modern Mandarin. Records contain phonetics, vocabulary, sentences, lessons. 249 page book contains all recorded text, translations, grammar, vocabulary, exercises. Best course on market. 3 12" 33⅓ monaural records, book, album.
98832-5 Set $12.50

CATALOGUE OF DOVER BOOKS

MANUAL OF THE TREES OF NORTH AMERICA, Charles S. Sargent. The basic survey of every native tree and tree-like shrub, 717 species in all. Extremely full descriptions, information on habitat, growth, locales, economics, etc. Necessary to every serious tree lover. Over 100 finding keys. 783 illustrations. Total of 986pp.
20277-1, 20278-X Pa., Two vol. set $9.00

BIRDS OF THE NEW YORK AREA, John Bull. Indispensable guide to more than 400 species within a hundred-mile radius of Manhattan. Information on range, status, breeding, migration, distribution trends, etc. Foreword by Roger Tory Peterson. 17 drawings; maps. 540pp.
23222-0 Pa. $6.00

THE SEA-BEACH AT EBB-TIDE, Augusta Foote Arnold. Identify hundreds of marine plants and animals: algae, seaweeds, squids, crabs, corals, etc. Descriptions cover food, life cycle, size, shape, habitat. Over 600 drawings. 490pp.
21949-6 Pa.$5.00

THE MOTH BOOK, William J. Holland. Identify more than 2,000 moths of North America. General information, precise species descriptions. 623 illustrations plus 48 color plates show almost all species, full size. 1968 edition. Still the basic book. Total of 551pp. 6½ x 9¼.
21948-8 Pa. $6.00

HOW INDIANS USE WILD PLANTS FOR FOOD, MEDICINE & CRAFTS, Frances Densmore. Smithsonian, Bureau of American Ethnology report presents wealth of material on nearly 200 plants used by Chippewas of Minnesota and Wisconsin. 33 plates plus 122pp. of text. 6⅛ x 9¼.
23019-8 Pa. $2.50

OLD NEW YORK IN EARLY PHOTOGRAPHS, edited by Mary Black. Your only chance to see New York City as it was 1853-1906, through 196 wonderful photographs from N.Y. Historical Society. Great Blizzard, Lincoln's funeral procession, great buildings. 228pp. 9 x 12.
22907-6 Pa. $6.95

THE AMERICAN REVOLUTION, A PICTURE SOURCEBOOK, John Grafton. Wonderful Bicentennial picture source, with 411 illustrations (contemporary and 19th century) showing battles, personalities, maps, events, flags, posters, soldier's life, ships, etc. all captioned and explained. A wonderful browsing book, supplement to other historical reading. 160pp. 9 x 12.
23226-3 Pa. $4.00

PERSONAL NARRATIVE OF A PILGRIMAGE TO AL-MADINAH AND MECCAH, Richard Burton. Great travel classic by remarkably colorful personality. Burton, disguised as a Moroccan, visited sacred shrines of Islam, narrowly escaping death. Wonderful observations of Islamic life, customs, personalities. 47 illustrations. Total of 959pp.
21217-3, 21218-1 Pa., Two vol. set$10.00

INCIDENTS OF TRAVEL IN CENTRAL AMERICA, CHIAPAS, AND YUCATAN, John L. Stephens. Almost single-handed discovery of Maya culture; exploration of ruined cities, monuments, temples; customs of Indians. 115 drawings. 892pp.
22404-X, 22405-8 Pa., Two vol. set $9.00

CONSTRUCTION OF AMERICAN FURNITURE TREASURES, Lester Margon. 344 detail drawings, complete text on constructing exact reproductions of 38 early American masterpieces: Hepplewhite sideboard, Duncan Phyfe drop-leaf table, mantel clock, gate-leg dining table, Pa. German cupboard, more. 38 plates. 54 photographs. 168pp. 8⅜ x 11¼. 23056-2 Pa. $4.00

JEWELRY MAKING AND DESIGN, Augustus F. Rose, Antonio Cirino. Professional secrets revealed in thorough, practical guide: tools, materials, processes; rings, brooches, chains, cast pieces, enamelling, setting stones, etc. Do not confuse with skimpy introductions: beginner can use, professional can learn from it. Over 200 illustrations. 306pp. 21750-7 Pa. $3.00

METALWORK AND ENAMELLING, Herbert Maryon. Generally conceded best all-around book. Countless trade secrets: materials, tools, soldering, filigree, setting, inlay, niello, repoussé, casting, polishing, etc. For beginner or expert. Author was foremost British expert. 330 illustrations. 335pp. 22702-2 Pa. $4.00

WEAVING WITH FOOT-POWER LOOMS, Edward F. Worst. Setting up a loom, beginning to weave, constructing equipment, using dyes, more, plus over 285 drafts of traditional patterns including Colonial and Swedish weaves. More than 200 other figures. For beginning and advanced. 275pp. 8¾ x 6⅜. 23064-3 Pa. $4.50

WEAVING A NAVAJO BLANKET, Gladys A. Reichard. Foremost anthropologist studied under Navajo women, reveals every step in process from wool, dyeing, spinning, setting up loom, designing, weaving. Much history, symbolism. With this book you could make one yourself. 97 illustrations. 222pp. 22992-0 Pa. $3.00

NATURAL DYES AND HOME DYEING, Rita J. Adrosko. Use natural ingredients: bark, flowers, leaves, lichens, insects etc. Over 135 specific recipes from historical sources for cotton, wool, other fabrics. Genuine premodern handicrafts. 12 illustrations. 160pp. 22688-3 Pa. $2.00

DRIED FLOWERS, Sarah Whitlock and Martha Rankin. Concise, clear, practical guide to dehydration, glycerinizing, pressing plant material, and more. Covers use of silica gel. 12 drawings. Originally titled "New Techniques with Dried Flowers." 32pp. 21802-3 Pa. $1.00

THOMAS NAST: CARTOONS AND ILLUSTRATIONS, with text by Thomas Nast St. Hill. Father of American political cartooning. Cartoons that destroyed Tweed Ring; inflation, free love, church and state; original Republican elephant and Democratic donkey; Santa Claus; more. 117 illustrations. 146pp. 9 x 12.
22983-1 Pa. $4.00
23067-8 Clothbd. $8.50

FREDERIC REMINGTON: 173 DRAWINGS AND ILLUSTRATIONS. Most famous of the Western artists, most responsible for our myths about the American West in its untamed days. Complete reprinting of *Drawings of Frederic Remington* (1897), plus other selections. 4 additional drawings in color on covers. 140pp. 9 x 12.
20714-5 Pa. **$5.00**

CATALOGUE OF DOVER BOOKS

How to Solve Chess Problems, Kenneth S. Howard. Practical suggestions on problem solving for very beginners. 58 two-move problems, 46 3-movers, 8 4-movers for practice, plus hints. 171pp.　　　　20748-X Pa. $3.00

A Guide to Fairy Chess, Anthony Dickins. 3-D chess, 4-D chess, chess on a cylindrical board, reflecting pieces that bounce off edges, cooperative chess, retrograde chess, maximummers, much more. Most based on work of great Dawson. Full handbook, 100 problems. 66pp. 7⅞ x 10¾.　　　22687-5 Pa. $2.00

Win at Backgammon, Millard Hopper. Best opening moves, running game, blocking game, back game, tables of odds, etc. Hopper makes the game clear enough for anyone to play, and win. 43 diagrams. 111pp.　　　22894-0 Pa. $1.50

Bidding a Bridge Hand, Terence Reese. Master player "thinks out loud" the binding of 75 hands that defy point count systems. Organized by bidding problem—no-fit situations, overbidding, underbidding, cueing your defense, etc. 254pp.　　　　　　　　　　　　　　　　　EBE 22830-4 Pa. $3.00

The Precision Bidding System in Bridge, C.C. Wei, edited by Alan Truscott. Inventor of precision bidding presents average hands and hands from actual play, including games from 1969 Bermuda Bowl where system emerged. 114 exercises. 116pp.　　　　　　　　　　　　　　　　　　　21171-1 Pa. $1.75

Learn Magic, Henry Hay. 20 simple, easy-to-follow lessons on magic for the new magician: illusions, card tricks, silks, sleights of hand, coin manipulations, escapes, and more —all with a minimum amount of equipment. Final chapter explains the great stage illusions. 92 illustrations. 285pp.　　　21238-6 Pa. $2.95

The New Magician's Manual, Walter B. Gibson. Step-by-step instructions and clear illustrations guide the novice in mastering 36 tricks; much equipment supplied on 16 pages of cut-out materials. 36 additional tricks. 64 illustrations. 159pp. 6⅝ x 10.　　　　　　　　　　　　　　　　23113-5 Pa. $3.00

Professional Magic for Amateurs, Walter B. Gibson. 50 easy, effective tricks used by professionals —cards, string, tumblers, handkerchiefs, mental magic, etc. 63 illustrations. 223pp.　　　　　　　　　　　　　23012-0 Pa. $2.50

Card Manipulations, Jean Hugard. Very rich collection of manipulations; has taught thousands of fine magicians tricks that are really workable, eye-catching. Easily followed, serious work. Over 200 illustrations. 163pp. 20539-8 Pa. $2.00

Abbott's Encyclopedia of Rope Tricks for Magicians, Stewart James. Complete reference book for amateur and professional magicians containing more than 150 tricks involving knots, penetrations, cut and restored rope, etc. 510 illustrations. Reprint of 3rd edition. 400pp.　　　　　　　　23206-9 Pa. $3.50

The Secrets of Houdini, J.C. Cannell. Classic study of Houdini's incredible magic, exposing closely-kept professional secrets and revealing, in general terms, the whole art of stage magic. 67 illustrations. 279pp.　　22913-0 Pa. $3.00

THE MAGIC MOVING PICTURE BOOK, Bliss, Sands & Co. The pictures in this book move! Volcanoes erupt, a house burns, a serpentine dancer wiggles her way through a number. By using a specially ruled acetate screen provided, you can obtain these and 15 other startling effects. Originally "The Motograph Moving Picture Book." 32pp. 8¼ x 11.                    23224-7 Pa. $1.75

STRING FIGURES AND HOW TO MAKE THEM, Caroline F. Jayne. Fullest, clearest instructions on string figures from around world: Eskimo, Navajo, Lapp, Europe, more. Cats cradle, moving spear, lightning, stars. Introduction by A.C. Haddon. 950 illustrations. 407pp.                    20152-X Pa. $3.50

PAPER FOLDING FOR BEGINNERS, William D. Murray and Francis J. Rigney. Clearest book on market for making origami sail boats, roosters, frogs that move legs, cups, bonbon boxes. 40 projects. More than 275 illustrations. Photographs. 94pp.
20713-7 Pa. $1.25

INDIAN SIGN LANGUAGE, William Tomkins. Over 525 signs developed by Sioux, Blackfoot, Cheyenne, Arapahoe and other tribes. Written instructions and diagrams: how to make words, construct sentences. Also 290 pictographs of Sioux and Ojibway tribes. 111pp. 6⅛ x 9¼.                    22029-X Pa. $1.75

BOOMERANGS: HOW TO MAKE AND THROW THEM, Bernard S. Mason. Easy to make and throw, dozens of designs: cross-stick, pinwheel, boomabird, tumblestick, Australian curved stick boomerang. Complete throwing instructions. All safe. 99pp.                    23028-7 Pa. $1.75

25 KITES THAT FLY, Leslie Hunt. Full, easy to follow instructions for kites made from inexpensive materials. Many novelties. Reeling, raising, designing your own. 70 illustrations. 110pp.                    22550-X Pa. $1.50

TRICKS AND GAMES ON THE POOL TABLE, Fred Herrmann. 79 tricks and games, some solitaires, some for 2 or more players, some competitive; mystifying shots and throws, unusual carom, tricks involving cork, coins, a hat, more. 77 figures. 95pp.                    21814-7 Pa. $1.50

WOODCRAFT AND CAMPING, Bernard S. Mason. How to make a quick emergency shelter, select woods that will burn immediately, make do with limited supplies, etc. Also making many things out of wood, rawhide, bark, at camp. Formerly titled Woodcraft. 295 illustrations. 580pp.                    21951-8 Pa. $4.00

AN INTRODUCTION TO CHESS MOVES AND TACTICS SIMPLY EXPLAINED, Leonard Barden. Informal intermediate introduction: reasons for moves, tactics, openings, traps, positional play, endgame. Isolates patterns. 102pp. USO 21210-6 Pa. $1.35

LASKER'S MANUAL OF CHESS, Dr. Emanuel Lasker. Great world champion offers very thorough coverage of all aspects of chess. Combinations, position play, openings, endgame, aesthetics of chess, philosophy of struggle, much more. Filled with analyzed games. 390pp.                    20640-8 Pa. $4.00

CATALOGUE OF DOVER BOOKS

SLEEPING BEAUTY, illustrated by Arthur Rackham. Perhaps the fullest, most delightful version ever, told by C.S. Evans. Rackham's best work. 49 illustrations. 110pp. 7⅞ x 10¾.                                                 22756-1 Pa. $2.00

THE WONDERFUL WIZARD OF OZ, L. Frank Baum. Facsimile in full color of America's finest children's classic. Introduction by Martin Gardner. 143 illustrations by W.W. Denslow. 267pp.                                              20691-2 Pa. $3.00

GOOPS AND HOW TO BE THEM, Gelett Burgess. Classic tongue-in-cheek masquerading as etiquette book. 87 verses, 170 cartoons as Goops demonstrate virtues of table manners, neatness, courtesy, more. 88pp. 6½ x 9¼.
22233-0 Pa. $2.00

THE BROWNIES, THEIR BOOK, Palmer Cox. Small as mice, cunning as foxes, exuberant, mischievous, Brownies go to zoo, toy shop, seashore, circus, more. 24 verse adventures. 266 illustrations. 144pp. 6⅝ x 9¼.     21265-3 Pa. $2.50

BILLY WHISKERS: THE AUTOBIOGRAPHY OF A GOAT, Frances Trego Montgomery. Escapades of that rambunctious goat. Favorite from turn of the century America. 24 illustrations. 259pp.                                           22345-0 Pa. $2.75

THE ROCKET BOOK, Peter Newell. Fritz, janitor's kid, sets off rocket in basement of apartment house; an ingenious hole punched through every page traces course of rocket. 22 duotone drawings, verses. 48pp. 6⅞ x 8⅜.        22044-3 Pa. $1.50

CUT AND COLOR PAPER MASKS, Michael Grater. Clowns, animals, funny faces . . . simply color them in, cut them out, and put them together, and you have 9 paper masks to play with and enjoy. Complete instructions. Assembled masks shown in full color on the covers. 32pp. 8¼ x 11.         23171-2 Pa. $1.50

THE TALE OF PETER RABBIT, Beatrix Potter. The inimitable Peter's terrifying adventure in Mr. McGregor's garden, with all 27 wonderful, full-color Potter illustrations. 55pp. 4¼ x 5½.                                    USO 22827-4 Pa. $1.00

THE TALE OF MRS. TIGGY-WINKLE, Beatrix Potter. Your child will love this story about a very special hedgehog and all 27 wonderful, full-color Potter illustrations. 57pp. 4¼ x 5½.                                    USO 20546-0 Pa. $1.00

THE TALE OF BENJAMIN BUNNY, Beatrix Potter. Peter Rabbit's cousin coaxes him back into Mr. McGregor's garden for a whole new set of adventures. A favorite with children. All 27 full-color illustrations. 59pp. 4¼ x 5½.
USO 21102-9 Pa. $1.00

THE MERRY ADVENTURES OF ROBIN HOOD, Howard Pyle. Facsimile of original (1883) edition, finest modern version of English outlaw's adventures. 23 illustrations by Pyle. 296pp. 6½ x 9¼.                                22043-5 Pa. $4.00

TWO LITTLE SAVAGES, Ernest Thompson Seton. Adventures of two boys who lived as Indians; explaining Indian ways, woodlore, pioneer methods. 293 illustrations. 286pp.                                                      20985-7 Pa. $3.00

HOUDINI ON MAGIC, Harold Houdini. Edited by Walter Gibson, Morris N. Young. How he escaped; exposés of fake spiritualists; instructions for eye-catching tricks; other fascinating material by and about greatest magician. 155 illustrations. 280pp. 20384-0 Pa. $2.75

HANDBOOK OF THE NUTRITIONAL CONTENTS OF FOOD, U.S. Dept. of Agriculture. Largest, most detailed source of food nutrition information ever prepared. Two mammoth tables: one measuring nutrients in 100 grams of edible portion; the other, in edible portion of 1 pound as purchased. Originally titled Composition of Foods. 190pp. 9 x 12. 21342-0 Pa. $4.00

COMPLETE GUIDE TO HOME CANNING, PRESERVING AND FREEZING, U.S. Dept. of Agriculture. Seven basic manuals with full instructions for jams and jellies; pickles and relishes; canning fruits, vegetables, meat; freezing anything. Really good recipes, exact instructions for optimal results. Save a fortune in food. 156 illustrations. 214pp. 6⅛ x 9¼. 22911-4 Pa. $2.50

THE BREAD TRAY, Louis P. De Gouy. Nearly every bread the cook could buy or make: bread sticks of Italy, fruit breads of Greece, glazed rolls of Vienna, everything from corn pone to croissants. Over 500 recipes altogether. including buns, rolls, muffins, scones, and more. 463pp. 23000-7 Pa. $4.00

CREATIVE HAMBURGER COOKERY, Louis P. De Gouy. 182 unusual recipes for casseroles, meat loaves and hamburgers that turn inexpensive ground meat into memorable main dishes: Arizona chili burgers, burger tamale pie, burger stew, burger corn loaf, burger wine loaf, and more. 120pp. 23001-5 Pa. $1.75

LONG ISLAND SEAFOOD COOKBOOK, J. George Frederick and Jean Joyce. Probably the best American seafood cookbook. Hundreds of recipes. 40 gourmet sauces, 123 recipes using oysters alone! All varieties of fish and seafood amply represented. 324pp. 22677-8 Pa. $3.50

THE EPICUREAN: A COMPLETE TREATISE OF ANALYTICAL AND PRACTICAL STUDIES IN THE CULINARY ART, Charles Ranhofer. Great modern classic. 3,500 recipes from master chef of Delmonico's, turn-of-the-century America's best restaurant. Also explained, many techniques known only to professional chefs. 775 illustrations. 1183pp. 6⅝ x 10. 22680-8 Clothbd. $22.50

THE AMERICAN WINE COOK BOOK, Ted Hatch. Over 700 recipes: old favorites livened up with wine plus many more: Czech fish soup, quince soup, sauce Perigueux, shrimp shortcake, filets Stroganoff, cordon bleu goulash, jambonneau, wine fruit cake, more. 314pp. 22796-0 Pa. $2.50

DELICIOUS VEGETARIAN COOKING, Ivan Baker. Close to 500 delicious and varied recipes: soups, main course dishes (pea, bean, lentil, cheese, vegetable, pasta, and egg dishes), savories, stews, whole-wheat breads and cakes, more. 168pp. USO 22834-7 Pa. $2.00

COOKIES FROM MANY LANDS, Josephine Perry. Crullers, oatmeal cookies, chaux au chocolate, English tea cakes, mandel kuchen, Sacher torte, Danish puff pastry, Swedish cookies —a mouth-watering collection of 223 recipes. 157pp.
22832-0 Pa. $2.25

ROSE RECIPES, Eleanour S. Rohde. How to make sauces, jellies, tarts, salads, pot-pourris, sweet bags, pomanders, perfumes from garden roses; all exact recipes. Century old favorites. 95pp.
22957-2 Pa. $1.75

"OSCAR" OF THE WALDORF'S COOKBOOK, Oscar Tschirky. Famous American chef reveals 3455 recipes that made Waldorf great; cream of French, German, American cooking, in all categories. Full instructions, easy home use. 1896 edition. 907pp. 6⅝ x 9⅜.
20790-0 Clothbd. $15.00

JAMS AND JELLIES, May Byron. Over 500 old-time recipes for delicious jams, jellies, marmalades, preserves, and many other items. Probably the largest jam and jelly book in print. Originally titled May Byron's Jam Book. 276pp.
USO 23130-5 Pa. $3.50

MUSHROOM RECIPES, André L. Simon. 110 recipes for everyday and special cooking. Champignons à la grecque, sole bonne femme, chicken liver croustades, more; 9 basic sauces, 13 ways of cooking mushrooms. 54pp.
USO 20913-X Pa. $1.25

THE BUCKEYE COOKBOOK, Buckeye Publishing Company. Over 1,000 easy-to-follow, traditional recipes from the American Midwest: bread (100 recipes alone), meat, game, jam, candy, cake, ice cream, and many other categories of cooking. 64 illustrations. From 1883 enlarged edition. 416pp.
23218-2 Pa. $4.00

TWENTY-TWO AUTHENTIC BANQUETS FROM INDIA, Robert H. Christie. Complete, easy-to-do recipes for almost 200 authentic Indian dishes assembled in 22 banquets. Arranged by region. Selected from Banquets of the Nations. 192pp.
23200-X Pa. $2.50

*Prices subject to change without notice.*
Available at your book dealer or write for free catalogue to Dept. GI, Dover Publications, Inc., 180 Varick St., N.Y., N.Y. 10014. Dover publishes more than 150 books each year on science, elementary and advanced mathematics, biology, music, art, literary history, social sciences and other areas.